AF859879

THE COMPUTER IN TRAINING AND DEVELOPMENT

# THE COMPUTER IN TRAINING AND DEVELOPMENT

*David Drumm*

D. Reidel Publishing Company
A Member of the Kluwer Academic Publishers Group
Dordrecht/Boston/Lancaster

International Human Resources Development Corporation • Boston

**Library of Congress Cataloging in Publication Data**

Drumm, David, 1948–
The computer in training and development.

Includes index.
1. Electronic digital computers. 2. Computer managed instruction. I. Title.
QA76.5.D737 1985 001.64 84-6707
ISBN-13: 978-90-277-1981-2 e-ISBN-13: 978-94-009-5359-8
DOI: 10.1007/978-94-009-5359-8

Published by D. Reidel Publishing Company
P.O. Box 17, 3300 AA Dordrecht, Holland in co-publication with IHRDC

Sold and distributed in North America by IHRDC

In all other countries, sold and distributed by Kluwer Academic Publishers Group, P.O. Box 322, 3300 AH Dordrecht, Holland

*To my family: my sons Jason and Christopher, but most of all, to my incredibly supportive wife, Marian Clark-Drumm, whose ongoing belief in me and my work sustained both of us through the many long and often late hours it took to write this.*

# Contents

# Preface

Seldom if ever in recent history has a single technological advance made such sweeping changes in our lives as has the advancement of the digital computer. As you probably are aware, the computer itself is not a brand-new device. What is new (and growing every day) is the percentage of the population that is using computers in business and personal environments.

The personal computer's popularity is growing by leaps and bounds primarily because of its availability in a small and relatively inexpensive package (several hundred dollars), placing it well within the reach of many families and small businesses. Software to perform financial analysis, solve inventory problems, or just have fun with is readily available and reasonably priced, thereby absolving inexperienced users from the burdens of programming and the technical complexities associated with it. The typical user in this scenario derives much of the benefit and experiences few of the disadvantages. Yet, as expected, the solving of a problem often creates problems, as will be discussed.

Training and development personnel, like most other professionals, have not been immune to the onslaught of personal and business computers. Computer-aided and computer-managed instructional techniques (i.e., CAI and CMI) are common phrases today in the realm of training and development.

In a recent impromptu and thus unscientific survey, I noted that of roughly 150 trainers who were contemplating purchase of a personal computer, few had actually done it, although the majority had acknowledged the benefits of having one. In many conversations I have had over the past three years with training and development professionals, experienced or otherwise, I've sensed two emerging and dichotomous viewpoints: (1) the acknowledgment of the benefits of the computer and its use, and (2) the reluctance to get involved with it.

This book was prompted, in part, by my desire to bring business and training professionals closer to a real "comfort level" with the science and technology surrounding the computer. It is hoped that "to know is to use," while recognizing clearly that the computer will not and cannot solve all training and development issues.

This book will discuss what a computer really is and what it is not. I will attempt to address and place in context the structures, language, and buzzwords which, in my experience, have confused and discouraged many people. You should remember from the start that this book will not make computer scientists or technologists of you. That in itself should be a comforting thought. Hopefully you'll realize that you don't need that level of education or background anyway to properly and successfully implement the computer in selected applications in training and development.

The first seven chapters of this text are designed for anyone who wants to establish a rudimentary knowledge of the computer (computer literacy). Part II (chapters 8–10) will press on to the implementation of the computer in training and development. We'll tackle some grassroots issues like "why bother?," and then move on to describe the terminology and interface of computers and training in chapter 8. In chapter 9, the *hows* and *whens*, *dos* and *don'ts* are discussed. Through examination of case studies, chapter 10 will look at the thought process involved in implementing a computer-based training model. A Glossary follows.

This book is designed under the concept of mastery learning. That is, you should understand the material covered in chapter 1 before proceeding to chapter 2 and so on. To help you in this area, specific and measurable objectives are stated at the beginning of each chapter. These objectives also act as a summary of concepts covered in the chapter. In addition, summaries are frequently provided, not only at the end of the chapters, but sometimes at selected locations within the chapters as well.

Both the objectives and summaries should act as self-check mechanisms for you. When you encounter a summary, read carefully the points that have already been covered. If you don't feel comfortable with what the summary says you should know, go back and reread the associated sections. After reading each chapter, check yourself again by going back to the objectives at the beginning of that chapter and making sure that you can do, state, define, etc., as indicated. Then you'll be ready to go on to the next chapter.

At the end of this text, you'll find a feedback form. Please fill it out and return it to me. This book was written only after many hours of thought were expended on planning the most logical way to explain the concepts involved. Still, I recognize that you, the reader, are the best source of feedback for identifying where this job could be done better. Let me know what problems, if any, you had as you progressed. Your thoughts and ideas will be incorporated into future editions to the best extent possible.

You may notice as you read that I am being very repetitive in some areas. Believe me, I'm not trying to make the book any longer than it has to be. I suspect, however, that a majority of those who read this text are truly beginning learners of computer technology. This book is oriented toward those individuals. What I am doing is simply trying to ensure that you understand concepts as we go. This will hopefully make your first exposure to computers a positive one and prompt further curiosity and investigation into this amazing science.

Thanks. I hope you benefit from and enjoy this experience.

*Dave Drumm*

*Special thanks go to Thomas D. Clasper, Training Consultant, whose significant contributions and efforts in the area of computer-based training are deeply appreciated.*

*I would like to thank Ron Nicosen, Ed.D., Ithaca College, and my niece Sue, for their encouragement.*

# PART I

# GETTING INVOLVED

# Objectives

At the conclusion of this chapter, you should be able to

1. identify your anxieties surrounding computers and their use
2. feel comfortable about wanting to learn more about computers
3. motivate yourself to read the remainder of this text with a positive attitude

# 1 A New Era for Training and Development Professionals

## EXAMINING OUR FEARS FIRST

It is indeed true that many professionals in training and development, as well as in business and industry, are afraid to face the reality that the computer is becoming a prevalent force in society today, directly involving more and more of us. The computer is a change agent that impacts not only on what we do, but particularly on how we do it.

In the 1940s the first computer, the ENIAC, became a reality. It was heavy (50 tons of hardware), and it was big, consuming 1500 square feet of floor space. It used vacuum tubes, like those in the early television sets, and generated a lot of heat. It was limited in scope by today's standards and therefore also limited in usage. But it *was* a computer. It was the first electronic device that could produce a variety of decisions and output given a variety of input conditions. Many people, therefore, thought it could think. It truly was the first of the first generation computers.

In the past forty years since the birth of the ENIAC, science and technology have advanced beyond some of our wildest dreams. New experiences of mankind, notably the space program of the sixties and early seventies, have fostered many advances that have touched every one of us. *Digital* has become a household word and surfaces everywhere: in the watches we wear, in the way we cook, and even in the toys our children play with and the programs and commercials they watch on TV. All in the past 40 years! Compounding this is the fact that the development of the computer has progressed at anything but a linear rate. Many of the technological advances realized and employed today have only become reality in the last five years.

The last decade alone has truly been an interesting one and has witnessed many advances in computer technology. Today, through advances made in microelectronics (i.e., making electronics smaller but doing the same job) and integrated circuit technology, we are able to buy a *real* computer at an affordable price. Some of them even fit in our pockets when not in use. Many of our cars now feature them.

We know that computers are out there but so far have been able to observe them from a "safe" distance. We haven't gotten involved with them because we haven't had to. But, the time has arrived when a passive acknowledgment of their existence won't be enough for many of us, and we know it. The anxiety level in many of us is already raised because we know it but we don't know anything about computers. As professionals in our work, we like to be prepared: for meetings, for class (a lot of us are going back to school these days), for "the next day." Yet, many of us don't feel prepared for this one. We see our friends buying and using personal computers. Many of our workplaces are rethinking procedures and methods and computerizing to save money.

Training and development people are being called on more and more by vendors with computer-based product lines and delivery tools that draw beautiful graphics (*graphics* not *pictures*—did you notice?). Now we worry. We are in the middle of a revolution and we are not prepared for it. We now have anxiety and procrastinate learning. The revolution, meanwhile, grows.

Before you continue through the rest of this volume, I need to make sure you are comfortable doing it. To do that, recognizing that the only way to solve a problem is to identify and face it, let's take a closer look at the anxieties that many of us are experiencing and then attempt to turn them into motivations for learning.

*Fear of the Unknown*

Most of us are wary of what we don't know, particularly if we have reason to believe that it will affect us at some point in the future. At the risk of sounding a bit trite, that's why we don't jump right in the water when we go swimming at the beach or pool. Rest assured, this text will not "push you off the dock." Hopefully you will progress at a pace that is satisfying and not confusing. As professionals, you have all learned: at school, at work, at home. You have the capability to learn something of computer technology too. To learn the basics is not difficult, nor is it incomprehensible. If it doesn't make sense to you now, it's probably because you never received a structured approach to it.

*The Personalization Concept*

Computers, unlike most other electronic entities, are capable of carrying on a dialogue with you provided that the computer is so programmed. They can print English text "by themselves." Computers are often referred to as "he," "she," "they," or as being like or possessing the traits of humans. Rarely are computers referred to as "it." This personalization concept alone has tended to intimidate many of us.

Many of the people I have talked to are intimidated. An interesting insight came recently from a professional woman in Pennsylvania, who said that a computer recently "asked" her a question. She was unable to type the answer. She knew the answer. She was familiar with the keyboard on the CRT (cathode ray tube), which is much like a typewriter keyboard. She just couldn't type the answer. It was her first direct encounter with a computer and she was intimidated by the computer's "personality." She is not alone.

Over the years, many of us involved with computer technology in one way or another have tended to personalize the nature of the computer, not because we wanted to confound the rest of you, but because in the circumstances it just seemed natural. Let's understand one thing from the start. Computers are not human. In fact, their "intelligence" is born of human thought. It is nurtured by humans, and it is changed by humans as well. Computers are but a tool to serve people.

This whole discussion may sound elementary but the fact remains that many of us have yet to face this simple reality. The truth of the matter is that the problems that a computer solves were first solved by humans, who devised the means by which to solve them and then "inserted" (by programming) these solutions into the computer. The computer then simply regurgitates the solution when told by humans to do so. The real power of a computer lies not in its "intelligence," but in its ability to render solutions *very quickly* to a myriad of problems, based upon its programming.

### *Prevalence of Computers*

Our anxieties are heightened by the ever-increasing prevalence of computers. Direct contact with computers, relegated to a select few ten years ago, is becoming commonplace in many locations and situations where it wasn't practical before. And no wonder. You can buy a complete computer system with a CRT (display), disc storage (for saving your work), and a printer for under $2000 today. Microminiaturization of electronic circuits used in computers have even reduced the price of computers themselves to well below that of the peripheral devices, such as printers and CRTs, needed to allow the computer to communicate with you and you to communicate with it. These dramatic cost reductions have prompted businesses to rethink methods and procedures that have been in place for years. They are beginning to recognize the tremendous cost effectiveness and impact the computer can have on their cost of goods manufactured and sold.

The demand throughout the nation and the world for business and personal computers has led to the creation of a fiercely competitive marketplace. About three years ago, an acquaintance of mine bought a popular brand of personal computer, which he is now using in his home, for $774. Today, that same computer is available for $299 and the $299 version has four times the memory that the $774 version had three years ago. I bought one last year for $499 and thought that I was getting the deal of a lifetime. A brand of personal computer that sold for about $100 last year is selling for $39 now (probably $19.95 by the time you read this). Figure 1.1 visualizes the dramatic cost reductions in computers realized through both technological and competitive forces, showing trends in both cost and usage over the past thirty years. It certainly illustrates the momentum that faces us.

## CONQUERING OUR FEARS

We all know as professionals that objectivity must be a necessary part of our lives in the way we think. We also know that it's not always easy to be objective. Nonetheless, let's examine some facts that should give us an objective place from which to start our study.

1. *Computers are not human.* They are not capable of self-initiated thought. They are capable of processing representations of thought very quickly.

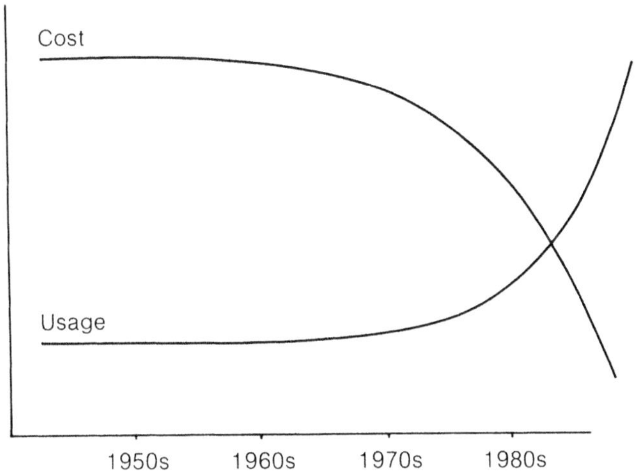


*Figure 1.1 Increasing prevalence of computers.*

2. *Computers are not smart.* They process data (numbers, etc.) as instructed by humans and in the order in which humans instruct them to process it. This yields results, as preconceived by humans, either visible or invisible.
3. *Computers are not infallible.* If the humans who program them make a mistake in their instructions, the computers will err in producing the expected results. Many of us have been affected by a computer error (i.e., *human* error).
4. Computers cannot always fix themselves if they fail to function normally. Highly qualified technicians (humans) are required to make a computer "healthy" again.
5. The study of computer science, in earnest, takes years of hard work in college or in a technical school. This not unlike other technical or professional disciplines that many of you have already mastered. Most of you will not ever become technicians in the field of computer technology. What's more important is that you don't have to become one to become an intelligent *user*, which many more of us are.

## WHY SHOULD I GET INVOLVED?

We already know that increased awareness and markedly decreased cost are factors in making the computer a popular option for many businesses. We can see it all around us. Your business may already have one or may be next. As a business professional, therefore, you need to become aware of those concepts that will allow you to further a more specific study as mandated by the application of computer technology in your home or office.

Training and development staff, investigating the implementation of computers for training, will need some basic concepts so that they can interact more intelligently with vendors of hardware and software. They may be chosen to aid in the evaluation of specific units. Which one of the many possible choices will do the job at the best price? This question embodies many concepts. Why let these concepts remain a mystery?

There are reasons indeed why you should get involved. A goal of any business is to optimize productivity, which contributes to profit. This is really the dawn of a new era. So, if you're ready, let's get started.

# Objectives

At the end of this chapter you should be able to

1. describe the use of DC voltage as the carrier of intelligence in a computer
2. describe how two DC voltages can be used to represent a number of different facts, states, or conditions
3. describe the basic structure of computer memory
4. count using the binary numbering system
5. define how memory is addressed
6. define the terms *bit* and *byte* and their relationship to a memory scheme
7. define an instruction
8. Define the CPU (central processing unit) of the computer and describe its functions
9. define the basic logic circuits used in a CPU, their symbols, and how they work
10. identify the truth tables for each logic circuit discussed
11. identify the outputs of simple logic configurations, given the inputs to those configurations

12. define the hardware components of a typical computer system and how the components communicate with each other

# 2 What a Computer Is—An Overview of Hardware

## INTRODUCTION

A computer is a configuration of electronic circuits which take combinations of voltages, representative of concepts, states, conditions, and facts, and produce other combinations of voltages which are representative of decisions, actions, or results. This is a good starting definition. Simply stated, a computer takes facts, conditions, and states in, which are represented by DC voltages, and produces results or decisions out. The facts, conditions and states are known as *input*. The results or decisions are known as *output*, as depicted in figure 2.1.

### *Voltages*

How do computers use voltages to process information? We already know that computers cannot think on their own. We know that the intelligence that they process is really ours (i.e., that of the person who programs the computer, known as a programmer). Since there is no convenient or viable way to directly insert a thought or series of thoughts into the computer for processing, a representative medium of intelligence must be used to allow the computer to function. This medium is known as *DC voltage*. DC (direct current) voltage is simply an electrical

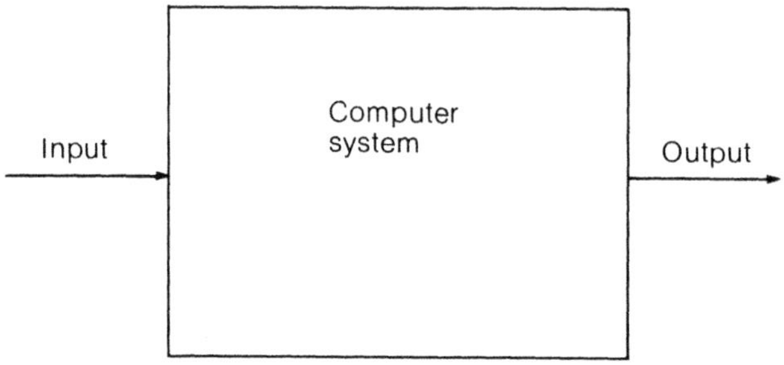


*Figure 2.1 Input conditions . . . output results.*

potential. Its direct application to electronic circuits allows them to work, as we will see in this chapter.

Think of the toys or portable appliances that you have around your home that require the use of batteries. When you insert the correct batteries and turn on the switch, the toy or appliance performs its designated function. It performs the same function every time you turn it on. The application of DC voltage, or power, to a computer allows it to work too, but unlike the toy or appliance the computer can perform many different functions. The functions it performs depend upon the input received. Remember that the input to computer circuits is in the form of DC voltage. As the input varies, so does the output, or results.

Not only is a digital computer reliant on DC power to work (i.e., on vs. off), it is also reliant on DC voltage to act as that "medium of intelligence" we referred to earlier. As you will discover, a computer processes unique combinations of DC voltage. Each combination will be known as an *instruction*. Each instruction causes different results to occur as the computer processes that instruction. Many instructions performed, or *executed*, in sequence, one at a time, allow the computer to produce the results that we see.

In figure 2.2 instructions are input to the computer. When they are executed, output or results are achieved. Instructions are input by a person known as a programmer.

### *Voltage Levels*

In the case of the toy or appliance that uses batteries, each battery has a specific voltage level, or potential, associated with

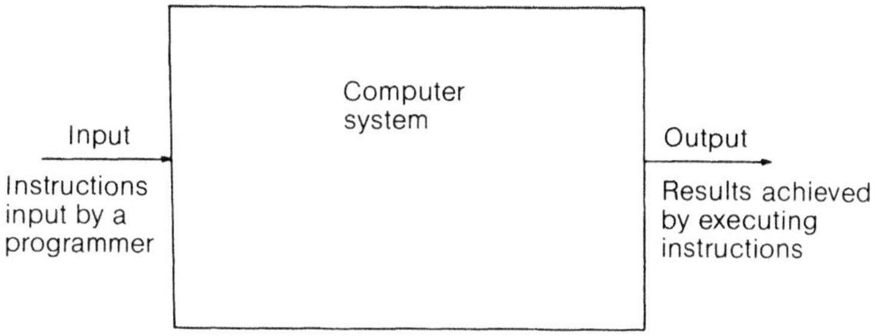


*Figure 2.2 Input instructions . . . output results.*

it. For instance we know that the large round "D" cells are rated at 1.5 volts DC each. The smaller rectangular batteries are each rated at 9 volts DC.

We now know that computers use voltage levels to represent intelligence, the presence or absence of certain facts, states, or conditions. Though not universal, a typical digital computer may use 5 volts DC and 0 volts DC as the voltage levels by which it maintains intelligence and makes decisions. The difference between the two obviously is that one is higher than the other. This is significant, as you will see.

At this point, it is not important that we understand everything there is to know about DC voltage and voltage levels. What is important to understand is that

- The electronic circuits of a digital computer use two different voltage levels to represent intelligence.
- The voltage levels used are DC voltages and one is higher than the other.

## REPRESENTING INTELLIGENCE

One immediately poses the question, How can the computer do all the things that it is capable of doing using only two voltages? The answer lies in the fact that a number of voltages, some high and some low, are combined together in unique combinations. The number of possible combinations produce many different possible instructions.

To see what we are dealing with, let's start out by representing two possible conditions: true and false. By assigning a voltage level of 5 volts DC to the true condition and 0 volts DC to the false condition, we are representing those two conditions in terms that a digital computer can "understand." This task of assignment is part of what a design engineer does as he designs the electronic circuitry that is the core of the computer itself.

Let's now represent the following:

- YES
- NO
- MAYBE

Right away, we recognize that there are three answers to be represented and only two DC voltage levels available to represent them. We now encounter the need to combine voltages together in some framework, as implied earlier. Figure 2.3 shows a possible solution.

What now becomes apparent is the addition and importance of some positional framework as well as the value of the voltage. Note carefully that the voltage values used to represent YES and NO are the same, but their positions are different, which makes the representation different.

Now, let's represent the following colors: blue, black, red, yellow, green, brown, purple, and orange. You can see the result in figure 2.4. Are all the combinations unique? Note in figure 2.4 that to represent eight colors, we had to expand to three positions in order to provide enough unique combinations.

| Representation | Meaning |
|---|---|
| 5 Volts; 0 Volts | Yes |
| 0 Volts; 5 Volts | No |
| 0 Volts; 0 Volts | Maybe |

*Figure 2.3 Representing three answers.*

| Representation (DC Volts) | | | |
|---|---|---|---|
| Position 3 | Position 2 | Position 1 | Meaning |
| 0 | 0 | 0 | Blue |
| 0 | 0 | 5 | Black |
| 0 | 5 | 0 | Red |
| 0 | 5 | 5 | Yellow |
| 5 | 0 | 0 | Green |
| 5 | 0 | 5 | Brown |
| 5 | 5 | 0 | Purple |
| 5 | 5 | 5 | Orange |

*Figure 2.4 Representing eight colors.*

## INTELLIGENCE IN COMPUTERS

Let's apply our introductory knowledge of representation techniques to its implementation in computers. First, recall that

- Instructions tell the computer what to do.
- Instructions are a unique combination of voltages that represent an action to be taken.
- The computer, upon receipt of an instruction, will execute it.

Now let's look at the hardware concepts surrounding a simple computer system. Using the idea of a positional framework introduced earlier, how many positions would be required such that 16 different instructions could exist? Referring to figure 2.5, we see that the answer is four.

Chapter 3 will examine some of these instructions, define what they actually do when executed, and implement them in a discussion and example of software principles. For now, however, we have discovered that instructions do exist, and are

| Representation (DC Voltage) | | | | |
|---|---|---|---|---|
| Position 4 | Position 3 | Position 2 | Position 1 | Meaning |
| 0 | 0 | 0 | 0 | Instruction #1 |
| 0 | 0 | 0 | 5 | Instruction #2 |
| 0 | 0 | 5 | 0 | Instruction #3 |
| 0 | 0 | 5 | 5 | Instruction #4 |
| 0 | 5 | 0 | 0 | Instruction #5 |
| 0 | 5 | 0 | 5 | Instruction #6 |
| 0 | 5 | 5 | 0 | Instruction #7 |
| 0 | 5 | 5 | 5 | Instruction #8 |
| 5 | 0 | 0 | 0 | Instruction #9 |
| 5 | 0 | 0 | 5 | Instruction #10 |
| 5 | 0 | 5 | 0 | Instruction #11 |
| 5 | 0 | 5 | 5 | Instruction #12 |
| 5 | 5 | 0 | 0 | Instruction #13 |
| 5 | 5 | 0 | 5 | Instruction #14 |
| 5 | 5 | 5 | 0 | Instruction #15 |
| 5 | 5 | 5 | 5 | Instruction #16 |

*Figure 2.5 Representing instructions.*

represented using combinations of DC voltages in some positional framework. Before we can implement them in a meaningful way, we have to examine some additional concepts in hardware.

### *Memory: A Way to Remember Instructions*

In the earliest computers, instructions had to be put in (input) to the computer one at a time and executed one at a time. There was no way for those early pioneers to "remember" or store many instructions at once. Today's computers can store, and thus "remember" many thousands of instructions and execute them sequentially at lightning speed by using a system component known as a memory.

A *memory* is nothing more than an "electronic warehouse" where thousands of instructions, as well as simple numbers, can be stored. The computer can look up the stored instructions one at a time and execute them, causing results or output to be achieved.

Memory takes the form of an electronic circuit board and varies in size depending upon the computer. It is often visualized as shown in figure 2.6. Memory consists of a number of *locations*. Each location can contain one instruction or one number. The number is often referred to as *data*.

### *Counting: The Computer's Way*

We know that conditions (such as true and false), answers (such as yes and no), and colors all can be represented using two DC voltages. In the case of colors, we combined individual voltages in a positional framework, where the position of each voltage had meaning too. One thing in common to all of this was what I will call the "rule of two," where only two voltages are ever used.

Because the rule of two applies universally in any digital computer, the computers of today use a different numbering system than most of us are aware of. That numbering system is called the *binary* numbering system. The binary numbering system has only two numbers, 0 and 1. This is contrasted, of course, to the decimal numbering system that we are all familiar with, which has ten numbers, 0–9, associated with it.

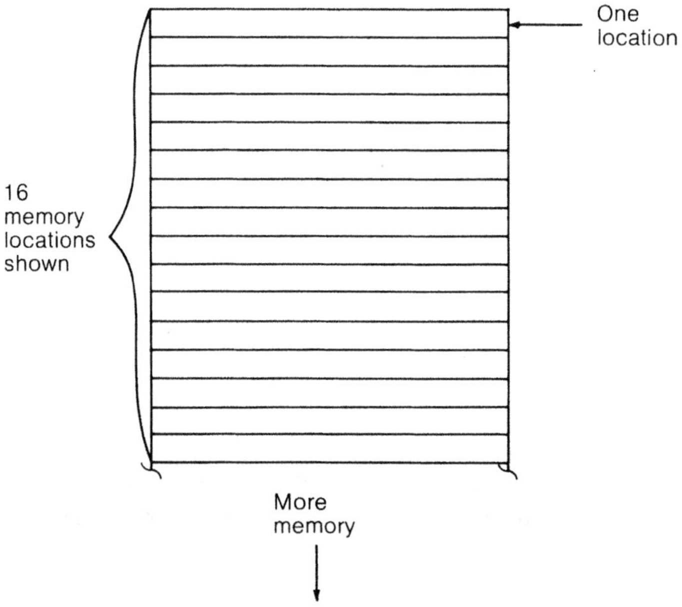


*Figure 2.6 Memory.*

*Counting in decimal.* What we are about to do does not imply that you don't know how to count in decimal. If, however, we take a look at how we count in decimal from a mechanistic point of view, the process of learning the binary number system will come much easier. Refer to figure 2.7. Count through the figure. Again, don't look at the numbers. Look at the mechanics.

Now refer to figure 2.8. In the positional framework here each position to the left has ten times the significance of its immediate position to the right.

*Counting in binary.* In the binary system only one thing changes. Since the binary number system has only two valid numbers instead of ten, each position to the left has twice the significance of its immediate position to the right. Now let's count. Refer to figure 2.9. Did you notice the same mechanics? Does this mean that with two values, we can indeed represent large numbers? Yes!

| Tens Column | Units Column | |
|---|---|---|
| 0 | 0 | |
| 0 | 1 | |
| 0 | 2 | |
| 0 | 3 | |
| 0 | 4 | |
| 0 | 5 | |
| 0 | 6 | |
| 0 | 7 | |
| 0 | 8 | |
| 0 | 9 | *Last* Valid Number |
| 1 | 0 | Return to *0* |
| | | Increment to *1* |

*Figure 2.7 Counting in decimals.*

Take the number 31, in both decimal and in binary, and examine it closely. Refer to figure 2.10. Large numbers *can* be represented in binary. The conclusion that we can draw from examination of figure 2.10 is that:

$$(11111)_2 = (31)_{10}$$

What is the binary number that represents the decimal number 212? Answer: (see fig. 2.11).

### *Equating Binary Numbers to Voltage Levels*

Since only two voltage levels are used in digital computers (the two we have discussed are 5 volts and 0 volts), and since the

| Hundreds | Tens | Units |
|---|---|---|
| 0 | 9 | 1 |
| 0 | 9 | 2 |
| 0 | 9 | 3 |
| 0 | 9 | 4 |
| 0 | 9 | 5 |
| 0 | 9 | 6 |
| 0 | 9 | 7 |
| 0 | 9 | 8 |
| 0 | 9 | 9 |
| 1 | 0 | 0 |

*Last* Valid Number

Increment To *1*

Return To *0*

*Figure 2.8 One more time . . .*

binary numbering system has only two valid numbers, 0 and 1, the two can be and are equated.

CONVENTION 1:

1 = 5 volts
0 = 0 volts

CONVENTION 2:

0 = 5 volts
1 = 0 volts

Both systems are used. The more common of the two is Convention 1, which we will implement in upcoming examples. This implies that the two DC voltage levels used in combination to comprise instructions can be directly related to the two values in our binary numbering system.

| Column | | | | |
|---|---|---|---|---|
| Sixteens | Eights | Fours | Twos | Ones |
| 0 | 0 | 0 | 0 | 0 |
| 0 | 0 | 0 | 0 | 1 |
| 0 | 0 | 0 | 1 | 0 |
| 0 | 0 | 0 | 1 | 1 |
| 0 | 0 | 1 | 0 | 0 |
| 0 | 0 | 1 | 0 | 1 |
| 0 | 0 | 1 | 1 | 0 |
| 0 | 0 | 1 | 1 | 1 |
| 0 | 1 | 0 | 0 | 0 |
| 0 | 1 | 0 | 0 | 1 |
| 0 | 1 | 0 | 1 | 0 |
| 0 | 1 | 0 | 1 | 1 |
| 0 | 1 | 1 | 0 | 0 |
| 0 | 1 | 1 | 0 | 1 |
| 0 | 1 | 1 | 1 | 0 |
| 0 | 1 | 1 | 1 | 1 |
| 1 | 0 | 0 | 0 | 0 |
| 1 | 0 | 0 | 0 | 1 |
| 1 | 0 | 0 | 1 | 0 |
| 1 | 0 | 0 | 1 | 1 |

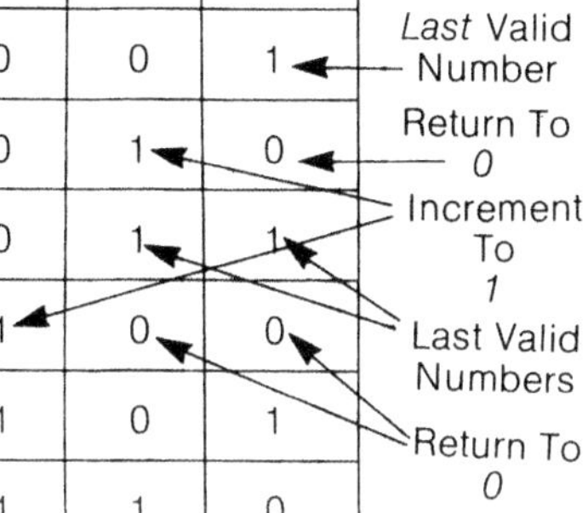


*Figure 2.9 Counting in binary.*

A. DECIMAL

| Tens | Units |
| --- | --- |
| 3 | 1 |

| Value | × Weight | Equals |
| --- | --- | --- |
| 3 | 10 | 30 |
| 1 | 1 | 1 |
| Total | | 31 |

B. BINARY

Position

| Sixteens | Eights | Fours | Twos | Ones |
| --- | --- | --- | --- | --- |
| 1 | 1 | 1 | 1 | 1 |

| Value | × Weight | Equals |
| --- | --- | --- |
| 1 | 16 | 16 |
| 1 | 8 | 8 |
| 1 | 4 | 4 |
| 1 | 2 | 2 |
| 1 | 1 | 1 |
| Total | | 31 |

*Figure 2.10 Decimal and binary representation.*

POSITION

| 128 | 64 | 32 | 16 | 8 | 4 | 2 | 1 |
|---|---|---|---|---|---|---|---|
| 1 | 1 | 0 | 1 | 0 | 1 | 0 | 0 |

$(11010100)_2$

PROOF

| Value | ×Weight | Equals |
|---|---|---|
| 1 | 128 | 128 |
| 1 | 64 | 64 |
| 0 | 32 | 0 |
| 1 | 16 | 16 |
| 0 | 8 | 0 |
| 1 | 4 | 4 |
| 0 | 2 | 0 |
| 0 | 1 | 0 |
| Total | | 212 |

*Figure 2.11* $(11010100)_2 = (212)_{10}$.

*Putting Concepts Together*

We now know

1. of the existence of a two-state system on which digital computers are based;
2. of the existence of the binary number system;
3. how to count in binary;
4. that a computer executes instructions that are a unique combination of binary digits (1 and 0), which translate for the computer's sake to 5 volts and 0 volts respectively.

Let's take those concepts and bring them together in a meaningful way.

*Binary/decimal equivalency.* We know that each binary number has a decimal equivalent. For the sake of clarity, let's look at both side by side. Refer to figure 2.12.

*Memory addresses.* Refer again to figure 2.6. That figure introduced the memory component and how it is visualized. Memory consists of a number of locations that are each used to store instructions or data. Since memory is the warehouse where all instructions and data are stored, and since memory can and often does consist of thousands of locations in total, a way to refer to each location must exist.

Each memory location has associated with it a number (a binary number) that serves the same purpose, such as the number on your house. Your house number uniquely identifies your house from every other house on your street. Your address identifies your location. The number associated with each memory location is known as an *address*. It serves to uniquely identify a specific memory location and set it apart from every other location in memory.

Let's now revisit figure 2.6 but this time assigning addresses to each memory location, starting at zero. Refer to figure 2.13.

*Bits—binary digits in memory.* The term *bits* is actually a contraction for *bi*nary digi*ts*. Figure 2.5 introduced the instruction set that we will use throughout chapters 2 and 3. Each instruction had four positions, which were required to allow definition of 16 different instructions.

Each position is known as a bit. The number of bits in a memory location defines the number of unique combinations that can be represented in that location. Now take figure 2.13 and add the bit concept. Refer to figure 2.14. Remember that each location will have four bits.

Note that

- each location has four bits
- each location has the same number of bits

All the bits in a single location form a *byte* (bite). That is, location (address) 0000 is a byte containing four bits.

16 Numbers Shown
As An Example

$(0000)_2$—$(1111)_2 \equiv (00)_{10}$—$(15)_{10}$

| Binary Number | | | | Decimal Equivalent | |
|---|---|---|---|---|---|
| 8s | 4s | 2s | 1s | 10s | 1s |
| 0 | 0 | 0 | 0 | 0 | 0 |
| 0 | 0 | 0 | 1 | 0 | 1 |
| 0 | 0 | 1 | 0 | 0 | 2 |
| 0 | 0 | 1 | 1 | 0 | 3 |
| 0 | 1 | 0 | 0 | 0 | 4 |
| 0 | 1 | 0 | 1 | 0 | 5 |
| 0 | 1 | 1 | 0 | 0 | 6 |
| 0 | 1 | 1 | 1 | 0 | 7 |
| 1 | 0 | 0 | 0 | 0 | 8 |
| 1 | 0 | 0 | 1 | 0 | 9 |
| 1 | 0 | 1 | 0 | 1 | 0 |
| 1 | 0 | 1 | 1 | 1 | 1 |
| 1 | 1 | 0 | 0 | 1 | 2 |
| 1 | 1 | 0 | 1 | 1 | 3 |
| 1 | 1 | 1 | 0 | 1 | 4 |
| 1 | 1 | 1 | 1 | 1 | 5 |

*Figure 2.12 Binary/decimal equivalency.*

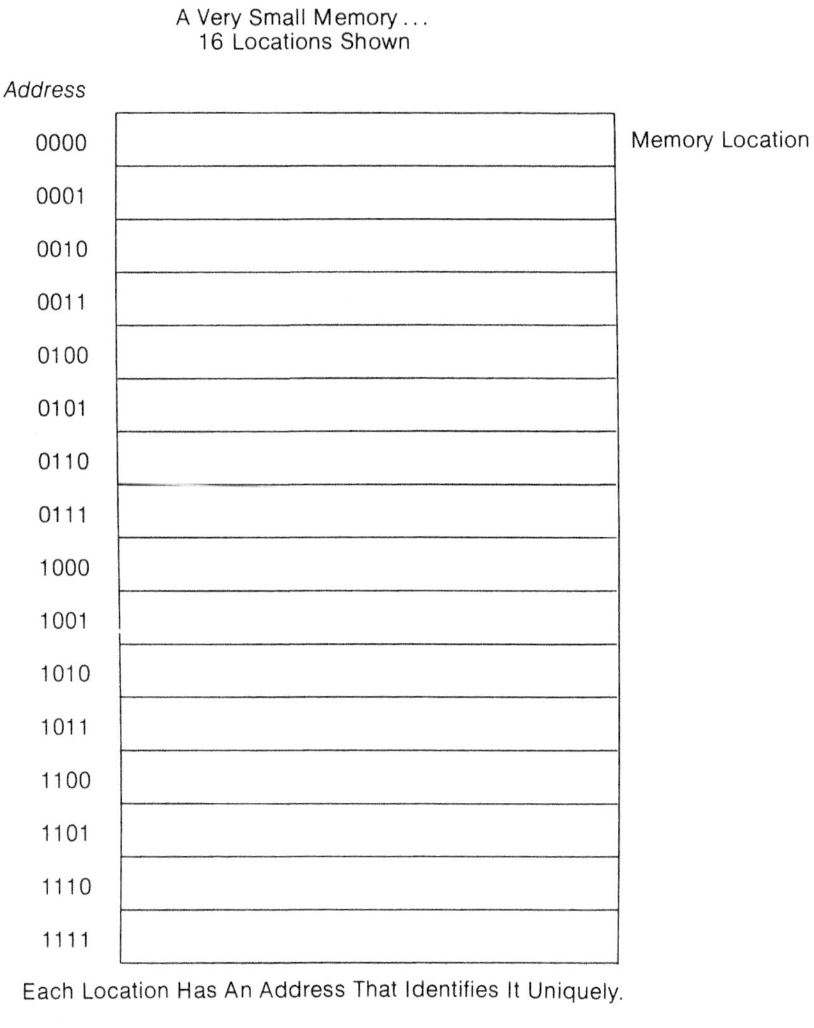


*Figure 2.13 Memory addressing.*

Since memory in a given computer always has a fixed byte size (all bytes have the same number of bits), memory dimensions are often referred to in terms of "length" and "width." The length is the number of bytes in memory. The width defines the number of bits per byte. Some examples follow:

$32 \times 8$ — 32 bytes of 8 bits each

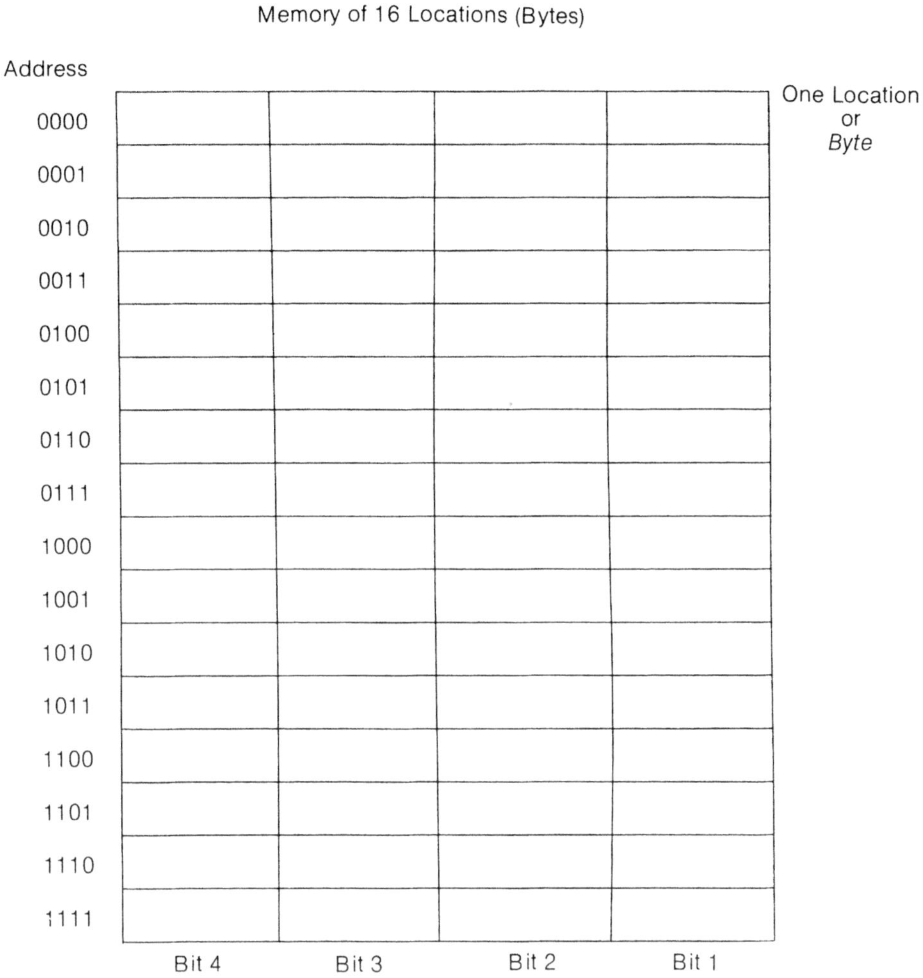


*Figure 2.14 Bits in memory.*

| | |
|---|---|
| 1024 × 8 | 1024 bytes of 8 bits each |
| 1024 × 16 | 1024 bytes of 16 bits each |

*Larger memories.* Memories in typical computers today are usually not small. Typical memory sizes in digital computers are 64K (65,536) bytes and larger. (1K is 1024 bytes; 64K, as it is often referred to, is 64 × 1024 bytes.)

Don't forget that just as memory locations themselves have a certain number of bits each, so do the associated memory addresses that identify those locations. In figure 2.14 the memory addresses for the partial memory shown each had four bits, not to be confused with the four bits of the location itself, which will be used to contain data and instructions.

For simplicity in previous examples and figures, a 4-bit memory and a very small memory size, or length, was used. Since the memory shown in figure 2.14 is limited to 16 bytes, addressing of those bytes can be accomplished with 4-bit addresses. Note that the addresses incremented from 0000 to 1111.

To address any more than 16 bytes, we would have to increase the size of our memory addressing scheme beyond four bits. Adding a fifth bit to our memory address would allow us to address 32 locations or twice the memory. This is because by using a 5-bit binary number we can count from 0 to 31 (32 unique numbers). This is shown in figure 2.15.

Try it and see. Count in binary from 00000 to 11111 using Worksheet 2.1. I've started you out.

Use of a 10-bit address would allow us to count to 1024 (1K) or address 1024 memory locations (2 to the 10th power = 1024). Use of a 16-bit address allows us to count to 65,536 (64K) or address 65,536 memory locations.

| Value | 16 | 8 | 4 | 2 | 1 | |
|---|---|---|---|---|---|---|
| Bit # | 5 | 4 | 3 | 2 | 1 | |
| All Bits 0 | 0 | 0 | 0 | 0 | 0 | Equals 0 |
| All Bits 1 | 1 | 1 | 1 | 1 | 1 | Equals 31 |

Range Is $(0\text{–}31)_{10}$
Or 32 Unique Numbers

*Figure 2.15 Binary/decimal equivalency.*

WORKSHEET 2.1.

| Binary | | | | | Decimal |
|---|---|---|---|---|---|
| 0 | 0 | 0 | 0 | 0 | 0 |
| 0 | 0 | 0 | 0 | 1 | 1 |
| 0 | 0 | 0 | 1 | 0 | 2 |
| 0 | 0 | 0 | 1 | 1 | 3 |
| | | | | | 4 |
| | | | | | 5 |
| | | | | | 6 |
| | | | | | 7 |
| | | | | | 8 |
| | | | | | 9 |
| | | | | | 10 |
| | | | | | 11 |
| | | | | | 12 |
| | | | | | 13 |
| | | | | | 14 |
| | | | | | 15 |
| | | | | | 16 |
| | | | | | 17 |
| | | | | | 18 |
| | | | | | 19 |
| | | | | | 20 |

*(continued on next page)*

WORKSHEET 2.1. *(continued)*

| Binary | | | | | Decimal |
|---|---|---|---|---|---|
| | | | | | 21 |
| | | | | | 22 |
| | | | | | 23 |
| | | | | | 24 |
| | | | | | 25 |
| | | | | | 26 |
| | | | | | 27 |
| | | | | | 28 |
| | | | | | 29 |
| | | | | | 30 |
| 1 | 1 | 1 | 1 | 1 | 31 |
| 16 | 8 | 4 | 2 | 1 | |

Bit Value

*Reviewing Bits, Bytes, and Memory Addresses*

First, some simple statements:

- Each memory location is a byte.
- Each byte can have a varying number of bits. Most microcomputers today have either 8 or 16 bits per location, or byte.
- Each byte has its own address, unique to it.
- The size of the address (i.e., the number of bits in the address) must be a function of the size (length) of the memory.

Ponder these examples:

1. A memory of 1024 8-bit bytes has
   - 8 bits per location (byte)

- 1024 locations
- an address length of 10 bits

2. A memory of 65,536 16-bit bytes has
   - 16 bits per byte
   - 65,536 bytes
   - an address length of 16 bits

## Conclusions:

1. The size of (number of bits in) the address is indicative of how many bytes can be directly addressed.
2. The size of the byte is indicative of how many different instructions can be stored and the size (magnitude) of numbers (data) that can be stored.

## MORE HARDWARE CONCEPTS

The memory component of a digital computer is not the only component. By examining it first we were able to look at some of the basic structures that surround a computer as well as the counting mechanisms that it uses. The memory is only used to store instructions that will be executed at some future time. They will not be executed by the memory, but by another major component called the central processing unit.

### *The Central Processing Unit (CPU)*

The central processing unit, or CPU, is that component of a digital computer that actually executes the instructions and manipulates the data stored in the memory component. It is the workhorse by which all results are achieved.

The CPU will receive an instruction from memory and examine that unique combination of "highs" and "lows" (1s and 0s) that it contains. It will examine the state (value) of each bit, take into consideration the position of each bit, and *decode* (figure out) what that instruction is. It will then cause the defined action associated with that instruction to take place. This process, known as *execution*, takes place for every instruction. *Logic circuits* (electronic circuits) provide the means by which instructions are decoded and executed.

*Logic Circuits*

We already know that a computer uses DC voltage levels to represent intelligence. Each of those levels is equated to a "1" or a "0," the two digits in the binary numbering system. This was done so that computer systems, which are nothing more than configurations of electronic circuits, could "understand" this intelligence and process it. We know that a CPU's job is to do the processing, to execute those instructions stored in memory. But how is it done?

The configuration of electronic circuits is actually comprised of many thousands of modules known as logic circuits. Logic circuits take one or more voltage inputs and produce one output dependent on what the input states are. There are several different logic circuits, which produce outputs under varying input conditions.

*The* AND *logic circuit.* The *AND* logic circuit is shown in figure 2.16. The *AND* logic circuit has a varying number of inputs (two are shown in the figure) and one output. It will always have one output.

RULE: The *AND* logic function, or *gate* as it is often referred to, will produce a high output if all of the inputs are high. If any input is low, the output will be low.

It is important to recognize that the symbol you see in figure 2.16 is just that, a symbol. It actually represents a combination of electronics which together perform the function indicated. Stated another way, the rule above says that if all inputs are at 5 volts, the output will be at 5 volts. If any input changes to 0 volts, the output will change to 0 volts.

Given an *AND* gate of two inputs, four possible combinations of inputs can be realized. They are defined in figure 2.17. The figure shows the outputs of the gate with the respective input combinations.

A note at this point. It is cumbersome to show actual voltages all the time. Also, as indicated earlier, the voltage levels used by a digital computer may not always be 5 volts and 0 volts. They will, however, always equate to binary numbers 0 and 1. For these reasons, tables such as the one shown in figure 2.17, which demonstrate the behavior of logic functions, will always be

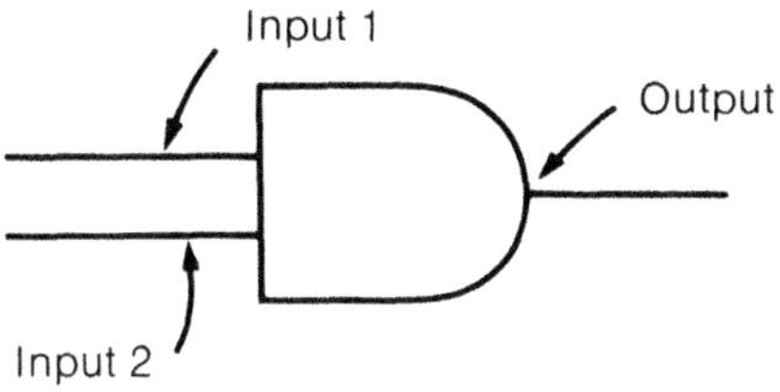


*Figure 2.16* AND *logic circuit.*

shown using the binary 0 and 1 representation instead. Draw the same table again, but this time using the binary equivalents. Refer to figure 2.18.

The table in figure 2.18 is referred to as a truth table. A truth table shows what the output of a logic circuit will be as the inputs change. What is important to recognize in the case of the *AND* logic circuit is that the output will only go high (5 volts DC), or go to a 1 state, when all of the inputs are high (5 volts DC).

Logic circuits can have a varying number of inputs. A three-input *AND* gate can have eight possible input combinations, but the only input combination that will yield a high, or 1, output is the combination where all inputs are high. Refer to figure 2.19. A three-input *AND* gate is shown with its associated truth table.

DC Volts

| If Input #1 Is | And Input #2 Is | Then The Output Is |
|---|---|---|
| 0 | 0 | 0 |
| 0 | 5 | 0 |
| 5 | 0 | 0 |
| 5 | 5 | 5 |

*Figure 2.17* AND *input and output behavior.*

| Input 1 | Input 2 | Output |
|---|---|---|
| 0 | 0 | 0 |
| 0 | 1 | 0 |
| 1 | 0 | 0 |
| 1 | 1 | 1 |

*Figure 2.18 Truth table for two input* AND *gate.*

*The* OR *logic gate circuit.* The *OR* gate is shown in figure 2.20 along with a truth table for two inputs.

RULE: The *OR* logic function will produce a high (5 volts DC) output if any of the inputs are high. If all of the inputs are low (0 volts DC), the output will be low.

*Inverter.* An inverter has only one input. Its job is to change the input state to the opposite state. It is shown in figure 2.21.

### *A Simple Example of Logic Combinations*

It is apparent that logic circuits such as those discussed in the previous sections do not do much by themselves. But logic gates in a digital computer never act alone. They are combined with other logic gates where the output of one gate becomes an input to another. By combining the functions of gates in this manner, many input combinations can be examined and acted upon. For example, the combined efforts of the eight logic functions shown in figure 2.22 allow us to look at eight different inputs and produce two outputs based upon the state of the inputs.

There are 256 (2 to the 8th power) possible input combinations in figure 2.22 and 4 (2 to the 2nd power) possible output combinations. To demonstrate how this circuit works, take a given set of inputs and derive the outputs. Given:

Input 1: 1(5 volts DC)
Input 2: 1(5 volts DC)

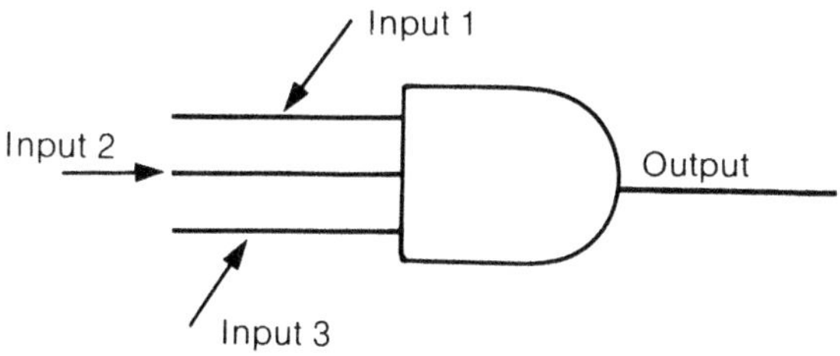


| Input 1 | Input 2 | Input 3 | Output |
|---|---|---|---|
| 0 | 0 | 0 | 0 |
| 0 | 0 | 1 | 0 |
| 0 | 1 | 0 | 0 |
| 0 | 1 | 1 | 0 |
| 1 | 0 | 0 | 0 |
| 1 | 0 | 1 | 0 |
| 1 | 1 | 0 | 0 |
| 1 | 1 | 1 | 1 |

*Figure 2.19 Three input* AND *gate and truth table.*

Input 3: 1(5 volts DC)
Input 4: 0(0 volts DC)
Input 5: 1(5 volts DC)
Input 6: 1(5 volts DC)
Input 7: 1(5 volts DC)
Input 8: 1(5 volts DC)

Given the inputs above, can you derive the outputs? It's not difficult. Just apply the simple rules for the logic gates you've studied thus far. Fill in your answers in the spaces provided and then read the analysis that follows to see how you did.

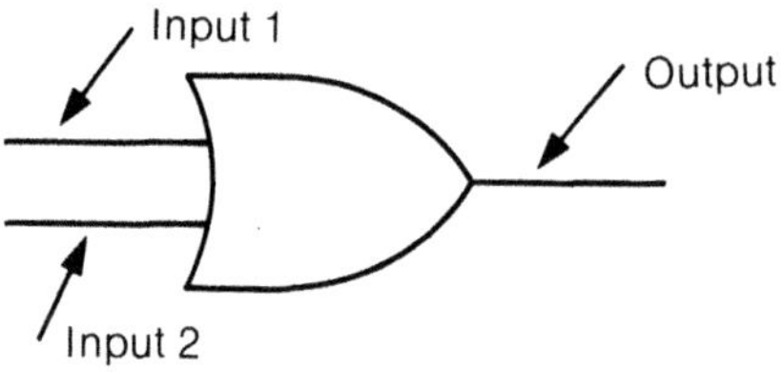


| Input 1 | Input 2 | Output |
|---|---|---|
| 0 | 0 | 0 |
| 0 | 1 | 1 |
| 1 | 0 | 1 |
| 1 | 1 | 1 |

*Figure 2.20 Two input* OR *gate and truth table.*

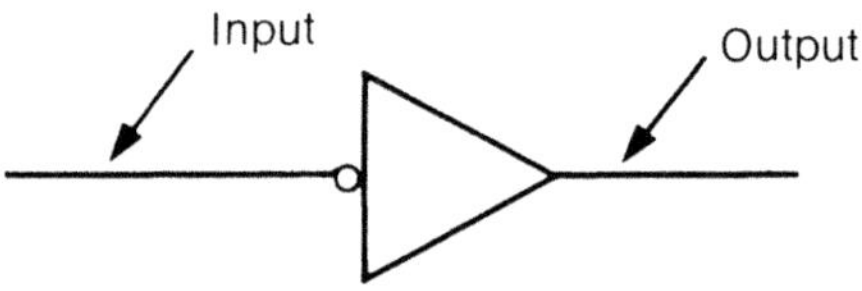


| Input | Output |
|---|---|
| 0 | 1 |
| 1 | 0 |

*Figure 2.21 Inverter with truth table.*

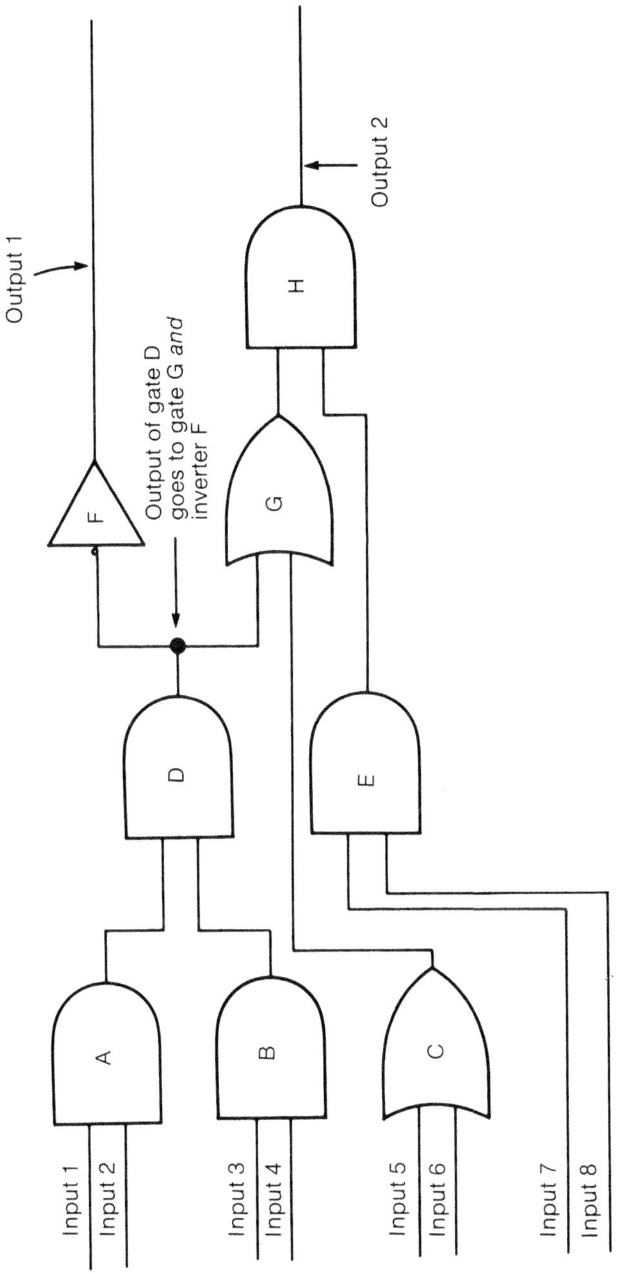


*Figure 2.22 A typical logic configuration.*

Output 1 = ____________

Output 2 = ____________

ANALYSIS: Refer to figure 2.22. Input 1 and 2 are high (1). The output of gate A is therefore high and becomes an input to gate D. Since input 4 is low (0), the output of gate B is low, and the second input to gate D is low. Gate D's output is therefore low, and inverted by Inverter F. Inverter F's output (Output 1) is high. Thus:

Output 1 = 1 (high; 5 volts DC)

Since input 5 is high, the output from gate C is high which becomes an input to gate G. Gate G produces a high output to the first input to gate H. Inputs 7 and 8 are both high. Gate E therefore sends a high output to the 2nd input of gate H. Since both inputs to gate H are high (1), the output from gate H is high. Therefore:

Output 2 = 1 (high; 5 volts DC)

Did you see all of that? If not, review it if you are keenly interested, but remember that this book is simply trying to establish an awareness, not make engineers of you. If you don't remember the nuts and bolts, it's not crucial, but hopefully you will gain some insight and appreciation for the mechanistic behavior of a computer. Really, it is nothing more than a "voltage processor."

*A Simple Exercise in Logic*

Let's look at a verbal logical process. I would like to get a quart of milk from the store. I process a series of thoughts related to that.

Is it raining? (yes or no)
If yes, do I have an umbrella? (yes or no)
If yes, then we'll go to the store.

If I don't have an umbrella, do I really need the milk? (yes or no)

If yes, then I go to the store anyway.
If no, then I stay home and wait for another day.

There is indeed a very logical flow of ideas and decisions. Refer to figure 2.23.

Could a combination of electronic circuits be designed to process that flow? What are the input states, facts, and conditions?

- whether or not it is raining
- whether or not I have an umbrella
- whether or not I really need the milk

What are the output decisions to be made?

- I will go to the store, or
- I will not go to the store

Remembering that voltages are simply a representation of facts, states, conditions, and decisions, could a circuit process my thoughts? Yes! Assign voltage levels to the inputs:

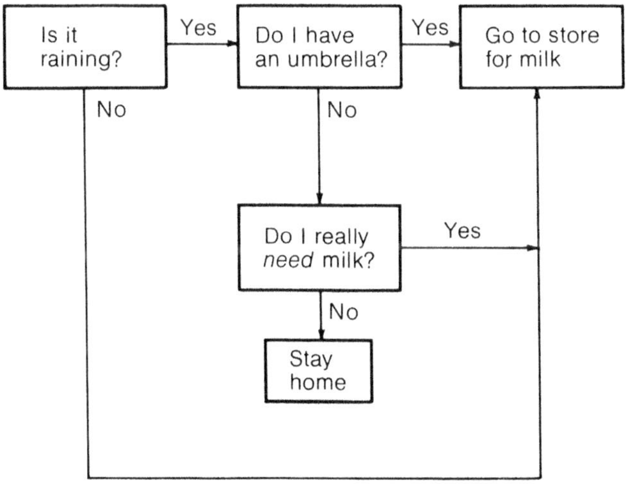


*Figure 2.23 A logical flow of ideas.*

raining = high (1)
not raining = low (0)

have umbrella = high (1)
no umbrella = low (0)

really need milk = high (1)
don't really need milk = low (0)

Next, assign voltage levels to the outputs:

go to store = high (1)
don't go to store = low (0)

Now refer to figure 2.24.

Given the circuit of logic gates shown in figure 2.24 and having assigned voltage levels to represent input conditions and output results or decisions, this circuit can take input states, facts, or conditions and derive output results or decisions. Examine several cases to see if it really works that way. Remember, the premise to start is that I need a quart of milk. If I did not need the milk to begin with, this example would be invalid.

CASE 1. It is raining and we have an umbrella; therefore we will go to the store. See figure 2.25.

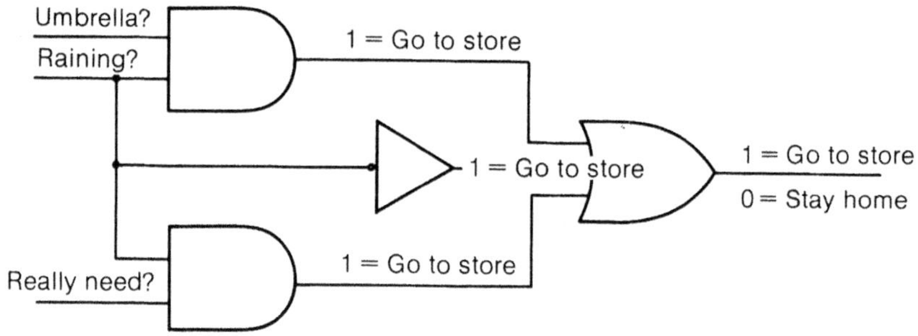


*Figure 2.24 Logic for milk example.*

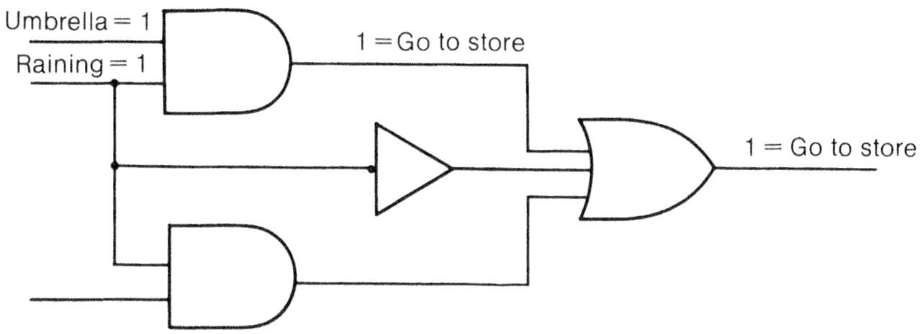


*Figure 2.25 Case 1.*

CASE 2. It is raining but we do not have an umbrella. We really need the milk. We will go to the store. See figure 2.26.

CASE 3. It is not raining, so we will go to the store. See figure 2.27.

Does it work? Yes! This example demonstrates the fact that we can process a logical flow of ideas if

- we can represent pertinent inputs and outputs with voltage levels

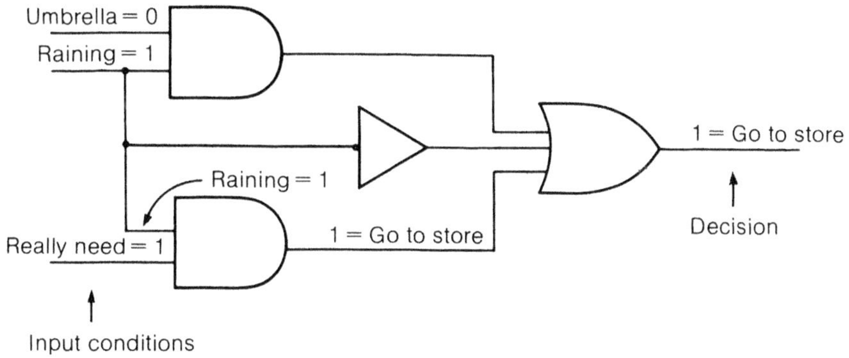


*Figure 2.26 Case 2.*

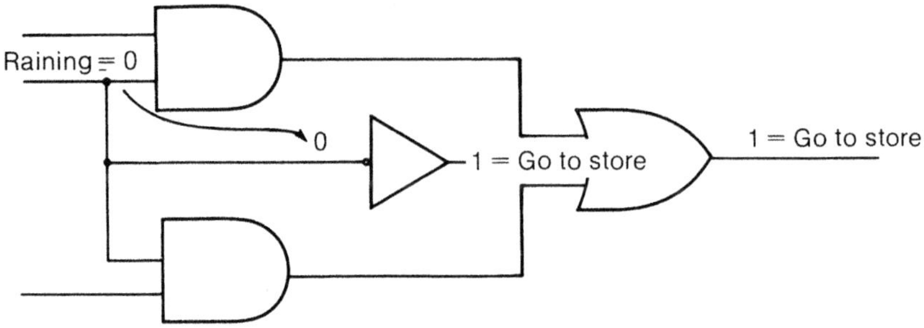


*Figure 2.27 Case 3.*

- we can connect circuitry such that the given inputs yield the desired output

Note that before the circuit became a reality, we designed it around known parameters to yield the desired results given the input.

In a manner of speaking, this circuit is a central processing unit. Compared to that of a computer, it processes very few input conditions, and yields very few possible results (only two total), but it is a type of central processing unit. Actual CPUs work this way using the basic rules set forth and the same type of logic circuits. The difference, of course, lies in the fact that real CPUs have the capability to examine, or decode, many more possible input combinations and generate more possible output results.

## A TYPICAL SYSTEM OF COMPUTER HARDWARE

Figure 2.28 shows and describes a typical computer system and its interconnections.

The basic elements of a computer's hardware are

1. central processing unit, as already described
2. memory, as already described

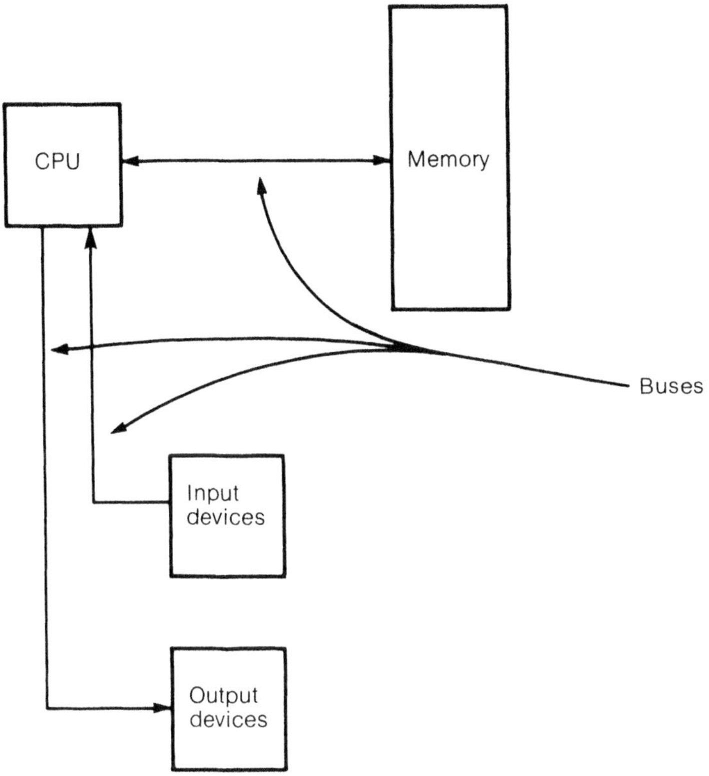


*Figure 2.28 A typical computer system.*

3. input devices, which allow the input of data to the computer from an external source
4. output devices, which allow the computer to communicate results or output to the user
5. buses, which are a number of wires over which information in the form of data or instructions (manifested as a combination of voltages) are sent from one point to another

Input and output devices are more fully described in chapter 5.

## SUMMARY

Computers use DC voltages to represent facts, states, or conditions (input) and yield results or decisions (output). Since two DC voltages are used, a numbering system with two values is used to represent values, count, and so forth.

Decisions are made using logic circuits, which take input voltages and produce output voltages, under a defined set of operating parameters. Major computer system components, such as the CPU, use many thousands of logic circuits to input instructions from memory, make decisions, and generate output based upon the input. Instructions are transported to the CPU from the memory as voltage combinations on a bus. The design (conception) of the circuits that make up a CPU is the responsibility of design engineers.

Instructions and data are stored in a memory. It is from the memory that instructions are extracted by the CPU, and executed, causing work to be accomplished. Instructions are a unique combination of voltages that represent an action to be taken. Input and output devices allow the user to input to the computer and the computer to output results or information to the user. Buses are employed here as well to allow transportation of data and instructions to and from the CPU.

The concept of hardware is probably not an easy one to grasp at first. You have been exposed to the concepts at an introductory level. Chapter 3 will use the hardware introduced here to explore the realm of software.

# Objectives

At the conclusion of this chapter, you should be able to

1. cite the factors that create the need to write a program
2. define a program
3. cite an example of the program-writing process
4. recognize and define terminology related to software
5. cite the consequences of incorrect instruction sequencing
6. cite the consequences of entering wrong instructions in a program

# 3 What a Computer Is—An Overview of Software

## INTRODUCTION

The discussion of software was actually begun in chapter 2. Software consists of the instructions stored in memory that, when executed by the CPU, cause work or results to be achieved. The instructions are grouped in a *program* and are defined in accordance with the problem that must be solved, or objectives that must be achieved.

## THE ENVIRONMENT AND CREATION OF SOFTWARE

### *The Problem—The Need*

Software is always created to solve a problem or satisfy a need. Software exists because of a need to solve an equation or send a string of characters to a printer, for example. All of the visible output from a computer that you see, such as reports on a printer or displays on a CRT, is the result of the execution of software.

A series of instructions is called a *program*. A program is *run*, or *executed*, any time the task or tasks accomplished by that program need to be performed. A program resides (is stored) in memory.

The longer a program is, in terms of the number of instructions that compose it, the more memory space is required to store it (i.e., more memory locations).

The program, once defined and written, is inserted into the computer's memory by a human being known as a programmer. The programmer's job is to take a known need and satisfy the need by writing a program. The following chain of events exemplifies the process of program definition:

- A problem or need is defined.
- The programmer writes a program to solve the problem or satisfy the need.
- The programmer enters the program into the memory of the computer.
- The programmer runs, or causes execution of, the program.

Quite often, programs do not initially satisfy the need or solve the problem. This is due to the existence of *bugs* or faults in the program. These bugs can occur for a number of different reasons and, depending on the nature of the bug, can cause a number of different and undesirable things to happen. A computer executes the instructions entered into its memory in the proper sequence when "told" to do so. If a wrong instruction is entered by the programmer or if the sequence of instructions is incorrect, an error will occur when the program is later executed. The results anticipated by the programmer will not be achieved. The programmer must find out why the program is not running correctly. Only when all bugs have been found and corrected will the program run smoothly and produce the desired results.

### *Writing a Program—The Logistics*

Programs are written and entered in many different ways depending on the *language* that is used to create the program. Every computer has a set of instructions, known as an *instruction set*, which its logic circuits (i.e., the logic circuits of the CPU) will recognize and execute. A programmer uses this instruction set to write a program in *machine language*, that is, in the language that the machine directly understands and can execute.

A reference manual describes all possible instructions and details the results when those instructions are executed by the computer. The programmer must define which instructions will be used and the sequence in which they should be executed. Once they are entered into the computer's memory via the keyboard, the programmer can cause the CPU to start execution of the program just entered.

In the simple example that follows, we will place ourselves in the shoes of the programmer tasked with solving a problem. This will create an awareness of the concerns and thought processes that must be dealt with in order to successfully write, enter, and implement a program.

*A Simple Example*

In chapter 2, figure 2.5 introduced the concept of instructions, each of which had four bits. Now that you have a basic grasp of the hardware concepts discussed there, let's involve ourselves in some programming concepts. What you will be exposed to is a simple example of the program-writing process. Most programs are longer and more complicated than this example. They also involve many technical concepts and abilities that are beyond the scope of this text.

Chapter 2 also introduced and discussed the concept of memory. An example shown in figure 2.14 was a 4-bit memory. Each memory location had 4 bits. First, recreate a portion of that memory in figure 3.1. The addresses for each location are numbered decimally (0–9) and in binary notation as well (0000–1001). During the following discussion memory will be referred to by decimal number for convenience. Figure 3.1 shows the memory of the "computer" that will be used in our example. Again, each memory location has four bits; there are a total of ten locations shown.

*The Problem and the Program*

THE PROBLEM: We need a short program to add two numbers and store the sum in a memory location. In a real environment we would store the sum in a memory location so it could be referenced and used later, perhaps by another program.

THE PROGRAM: As programmers, we must know where the two numbers to be added will come from. They could come from

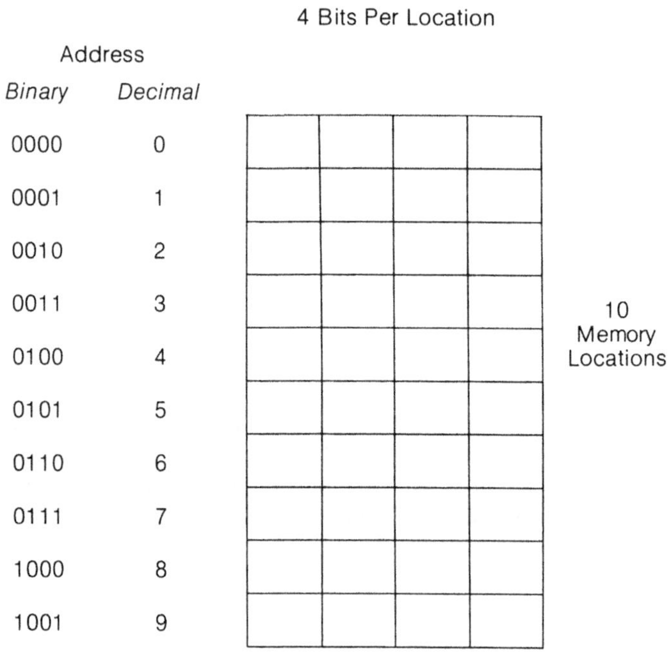


*Figure 3.1 Memory re-creation.*

- an input device (someone will type them in)
- other memory locations (they were either entered in those locations by the programmer or they were placed there as a result of execution of another program)

We will assume that the two numbers to be added will already exist in other memory locations. We could have entered them in those locations prior to running the program. They will be located in locations 6 and 7.

We must determine where the sum will be stored when addition is complete. We could pick any location, but for illustration's sake we will use location 8.

We know we must put the program in memory. It is from there that all programs are executed. Where shall we store this program? It doesn't really matter as long as we don't use locations 6, 7, or 8 as they are already allocated. We could start the program at location 0 and go from there as long as the length of the program does not exceed six locations, or instructions.

Now, on paper first, we define the tasks that must be done as well as the sequence of those tasks so that the goal of the program may be accomplished. The goal was to add two numbers, located in memory at locations 6 and 7, and store the sum in location 8.

1. Adding cannot be done directly in memory, but is done in the CPU. The arithmetic functions of a CPU, performed in the Arithmetic/logic unit (ALU), include addition, subtraction, multiplication, and division. In order to add, the CPU must first be in possession of the numbers to be added. Therefore, the first operation our program must perform is to load these numbers into the CPU from memory.
2. Now that the numbers have been loaded, the next step is to add them together.
3. The sum we now have in the CPU is never automatically stored in memory. It must be stored there using a separate instruction.
4. After implementing the instruction to store the sum, our work is complete. We must now tell the computer to stop with an instruction often referred to as a *halt* instruction.

A summary of the steps identified above is shown in figure 3.2.

Now we have to pick instructions from the instruction set (located in the reference manual) that will perform the steps we

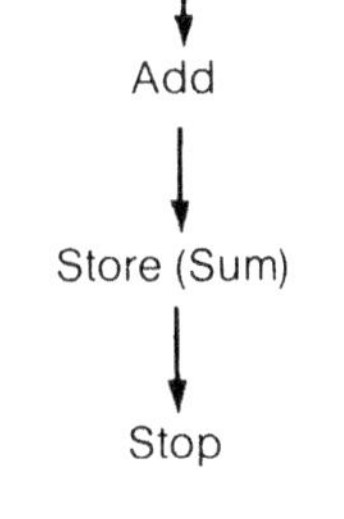


*Figure 3.2 Summary of program steps.*

have documented above. We also have to be very careful to sequence the instructions correctly. Then, it is simply a matter of loading those instructions into the computer's memory, in the proper sequence, and executing the resulting program.

First, to pick the proper instructions. In chapter 2, figure 2.5 listed a typical instruction set to show that they are nothing more than a unique combination of voltages (highs and lows). We did not define then what each instruction would do. Select eight of those 16 instructions and define now what each of them does. (See fig. 3.3.) What each instruction does is predefined by the manufacturer of the computer and is based upon what will happen when a particular instruction is executed.

Pick from figure 3.3 those instructions required to perform the job documented in figure 3.2. These are shown in figure 3.4. Note that the sequence of instructions matches the sequence of events in figure 3.2.

Since the series of instructions (program) in figure 3.4 matches the sequence defined in figure 3.2, it appears to be a workable program. The instructions in the "instruction" column will ultimately be entered into the computer's main memory prior to execution of the program.

Now let's enter the program, which we have decided will go into locations 0–4. Since there are five instructions in the program, five memory locations will be required to store these instructions in memory. The exact mechanics required to take a defined program and enter it into computer memory vary with just about every computer model made. However, most computers have this much in common for their program entry procedure:

1. A counting device, known generally as a program counter, that keeps track of which memory location is being accessed at any one point in time, is oriented (set) by the programmer to the memory address into which the first instruction will be entered. We decided our program counter would be set to location 0.
2. The first instruction is entered. In our example (see fig. 3.4) the first instruction is 0011, "load the contents of location 6 into the CPU." It is now placed in location 0.
3. The program counter is now incremented (increased in value by 1). Its value is now 1. Note: some computers increment the program counter automatically as instructions are entered. If this is not the case the programmer will have to

| Instructions | | | | Stored in Memory As (Voltage) | | | | Meaning* |
|---|---|---|---|---|---|---|---|---|
| Bit 4 | Bit 3 | Bit 2 | Bit 1 | Bit 4 | Bit 3 | Bit 2 | Bit 1 | |
| 0 | 0 | 1 | 1 | 0 | 0 | 5 | 5 | Load Contents of Location 6 Into CPU |
| 0 | 1 | 0 | 1 | 0 | 5 | 0 | 5 | Load Contents of Location 7 Into CPU |
| 0 | 0 | 0 | 0 | 0 | 0 | 0 | 0 | Stop (Halt) |
| 1 | 0 | 1 | 1 | 5 | 0 | 5 | 5 | Store Result In Location 8 |
| 1 | 0 | 0 | 0 | 5 | 0 | 0 | 0 | Add |
| 1 | 1 | 0 | 1 | 5 | 5 | 0 | 5 | Subtract 1st Number Loaded From 2nd Number Loaded |
| 0 | 1 | 1 | 0 | 0 | 5 | 5 | 0 | Load Contents of Location 9 Into CPU |
| 1 | 1 | 1 | 0 | 5 | 5 | 5 | 0 | Multiply 2 Numbers Loaded |

*What Will Happen When This Instruction Is Decoded and Executed by the CPU.

*Figure 3.3 "Reference manual."*

| Instruction | Meaning |
|---|---|
| 0011 | Load Contents of Location 6 Into CPU |
| 0101 | Load Contents of Location 7 Into CPU |
| 1000 | Add |
| 1011 | Store Sum In Location 8 |
| 0000 | Stop (Halt) |

*Figure 3.4 The program.*

manually increment the program counter prior to entering the next instruction.

4. The second instruction is entered (0101) and stored in location 1 in memory.
5. The program counter is incremented again. Its value is now at 2.
6. The third instruction (ADD) is entered and stored in location 2.
7. This process continues until all instructions in the program have been entered and placed in memory.

At this point our computer memory should contain the information shown in figure 3.5.

Partial Memory Shown

| Address | Bit 4 | Bit 3 | Bit 2 | Bit 1 |
|---|---|---|---|---|
| 0 | 0 | 0 | 1 | 1 |
| 1 | 0 | 1 | 0 | 1 |
| 2 | 1 | 0 | 0 | 0 |
| 3 | 1 | 0 | 1 | 1 |
| 4 | 0 | 0 | 0 | 0 |

*Figure 3.5 Program in memory.*

The program has been entered into memory. The data to be used by the program has not yet been entered. Using the same mechanical process we used to enter the program we must now enter the data. Knowing that the numbers stored in locations 6 and 7 will be added when the program is executed later, we must put into memory, at locations 6 and 7, the numbers we want to add. Thus:

1. Orient the program counter to location 6.
2. Enter the first number (7 in our example).
3. Increment the program counter (to location 7).
4. Enter another number (8 in our example).

At this point memory should appear as shown in figure 3.6.

Note that:

- There is a program area defined that contains the program.
- There is a data area defined that contains the data to be used by the program.

| Address | Bit 4 | Bit 3 | Bit 2 | Bit 1 | |
|---|---|---|---|---|---|
| 0 | 0 | 0 | 1 | 1 | |
| 1 | 0 | 1 | 0 | 1 | |
| 2 | 1 | 0 | 0 | 0 | Program Area |
| 3 | 1 | 0 | 1 | 1 | |
| 4 | 0 | 0 | 0 | 0 | |
| 5 | | | | | |
| 6 | 0 | 1 | 1 | 1 | Data Area |
| 7 | 1 | 0 | 0 | 0 | |
| 8 | | | | | |

*Figure 3.6 Program and data.*

- Location 8, which will be used to store the sum, has been shown blank. At this point, since we have only entered the program and have not as yet executed it, no adding has taken place. After we run the program once, location 8 should contain the sum of the two numbers now stored in locations 6 and 7.
- Location 5 is not used for anything (no special reason).
- The contents of locations 2 and 7 are the same—yet they are different indeed. The difference lies in the fact that the contents of location 2 will be interpreted as an *instruction* when executed. The contents of location 7 will be viewed as *data* by the program in locations 0–4. This points out the fact that data and instructions can have exactly the same binary value, but because of the way they are used, the meaning is entirely different. The value is the same but the significance is different to the computer. Because the contents of location 2 are part of a program, it will be treated as an instruction by the computer at execution time. This instruction in location 2, which "says" add, will treat the contents of location 7 as data.

*Executing the program.* Discussion and procedures thus far have

- defined the problem
- defined the nature and sequence of tasks needed to solve the problem
- defined the program (series of instructions) needed to solve the problem
- entered the program, and associated data required by the program, into the computer memory

Now let's execute the program.

The program counter defined earlier will now be used to indicate to the computer where the *origin* of the program is in order to begin execution. The origin of our program is location 0, since that is the location where the first instruction in the program was entered. To start the computer at any other point other than the origin would mean that only part of the program would be executed. Obviously, this is unsatisfactory.

Starting the computer to execute a program is again a function of button-pushing mechanics, which vary from computer to computer. Once the computer has been started, however, the following sequence typifies the process of program execution:

1. The instruction at the starting memory location is retrieved. It will be transported down the bus (see fig. 2.28) between the memory and the CPU, from the memory to the CPU.
2. The CPU decodes the bit combination (voltage combination) it receives and causes execution of the work defined by that instruction, using its logic circuits.
3. The program counter is incremented after the instruction is executed. This is an automatic procedure when programs are being executed.
4. The next instruction is retrieved from the next memory location.
5. This process is repeated.
6. The CPU finally encounters the stop instruction (location 4). The program has been executed as entered.

The above sequence showed what the computer does to execute a program. The following sequence attempts to integrate the actions of the programmer with the actions of the computer. You will observe repetition here but it is important to the goals of our experience to visualize, as best we can without the benefits of any hands-on time, what the interaction is. You should keep figure 3.6 handy as we progress through the following sequence.

ACTION: Programmer orients the program counter to location 0 which is the start of our program.

RESULT/MEANING: Computer is "in position" to start executing our program.

ACTION: Programmer starts computer running.

RESULT/MEANING: Execution begins.

ACTION: Computer retrieves instruction from location 0 and executes it.

RESULT/MEANING: Contents of location 6 (0111) are loaded into the CPU.

ACTION: Program counter is incremented automatically to location 1.

RESULT/MEANING: Computer is done executing the instruction at location 0. It is now ready to execute the instruction at location 1.

ACTION: Computer retrieves instruction from location 1 and executes it. (0101 is retrieved, decoded, and executed.)

RESULT/MEANING: Contents of location 7 (1000) are loaded into the CPU.

ACTION: Program counter is incremented automatically to location 2.

RESULT/MEANING: Computer is finished executing the instruction at location 1. It is now ready to execute the instruction at location 2.

ACTION: Computer retrieves instruction from location 2 and executes it. (1000 is retrieved, decoded, and executed.)

RESULT/MEANING: The CPU adds the two numbers it now has (0111 and 1000) and temporarily holds the sum.

ACTION: Program counter is incremented.

RESULT/MEANING: Computer is finished executing the instruction at location 2. It is now ready to go on to location 3.

ACTION: Computer retrieves instruction from location 3 and executes it. (1011 is retrieved, decoded, and executed.)

RESULT/MEANING: The CPU now takes the sum it was temporarily holding and stores it in location 8 in memory.

ACTION: Program counter is incremented.

RESULT/MEANING: Computer is finished executing the instruction at location 3 and is now ready to go on to location 4.

ACTION: Computer retrieves the instruction from location 4 and executes it. (0000 is retrieved, decoded, and executed.)

RESULT/MEANING: The CPU stops!

We have now executed the program. Look at figure 3.7. Everything was as it was prior to program execution except that the sum of locations 6 and 7 is now stored in location 8.

| Address | | | | | |
|---|---|---|---|---|---|
| 0 | 0 | 0 | 1 | 1 | |
| 1 | 0 | 1 | 0 | 1 | |
| 2 | 1 | 0 | 0 | 0 | |
| 3 | 1 | 0 | 1 | 1 | |
| 4 | 0 | 0 | 0 | 0 | |
| 5 | | | | | |
| 6 | 0 | 1 | 1 | 1 | |
| 7 | 1 | 0 | 0 | 0 | |
| 8 | 1 | 1 | 1 | 1 | ← Sum |

*Figure 3.7 Program executed.*

### *Variations on a Theme*

We have taken quite a bit of time to elaborate on an example of program concept, program writing, entering, and execution. Now, because you have reached a very basic level of technical maturity, let's play "what if?" in order to demonstrate that a computer is truly not a thinking, creative, or smart being but rather a very mechanistic tool.

#### *Variation 1: Incorrect sequencing of instructions*

SITUATION: The programmer has correctly defined the program on paper, as shown earlier in figure 3.4, but enters it into memory incorrectly. After program entry, memory looks like that shown in figure 3.8, with the stop instruction entered in location 2 instead of location 4, where it should be.

After entering the data in locations 6 and 7, and setting up the computer for execution of the program, the programmer starts the computer in execution. Will the sum be calculated and

| Address | | | | | |
|---|---|---|---|---|---|
| 0 | 0 | 0 | 1 | 1 | |
| 1 | 0 | 1 | 0 | 1 | |
| 2 | 0 | 0 | 0 | 0 | Misplaced Instruction |
| 3 | | | | | |
| 4 | | | | | |

*Figure 3.8 Incorrect sequence of instructions.*

stored in location 8? The answer, of course, is no. The program was started at location 0, as before, but when it executes the instruction at location 2 it will stop. No principles have changed, but because the stop instruction was misplaced, the program will not function as intended. The misplacement in this example is very obvious since the program is so short; however, it demonstrates the point.

*Variation 2: Incorrect instruction*

SITUATION: Instead of an instruction to add (1000), the programmer mistakenly selects an instruction to subtract the two numbers (1101; see fig. 3.3) and enters that in place of the add instruction. Thus, after entering the program into memory, memory looks like that shown in figure 3.9.

The programmer now enters the data in locations 6 and 7, and starts the program in execution. After the program has been executed, location 8, where the sum of the two numbers was supposed to be stored, shows the number 1 stored instead of 15. (We know why.) At this point the programmer is baffled. To find the error the program is run again, first entering two different numbers in the data locations (6 and 7). A 2 is entered in location 6 and a 5 in location 7. The program is run again and a 3 appears in location 8 at the end of program execution.

*Debugging.* The programmer has a problem. The program does not add correctly. It must be debugged by correcting the error. In the case of Variation 2, of course, the error is corrected by

| Address | | | | | |
|---|---|---|---|---|---|
| 0 | 0 | 0 | 1 | 1 | |
| 1 | 0 | 1 | 0 | 1 | |
| 2 | 1 | 1 | 0 | 1 | ← Subtract |
| 3 | 1 | 0 | 1 | 1 | |
| 4 | 0 | 0 | 0 | 0 | |
| 5 | | | | | |
| 6 | 0 | 1 | 1 | 1 | |
| 7 | 1 | 0 | 0 | 0 | |
| 8 | 0 | 0 | 0 | 1 | ← 1 Instead of $15_{(1111)}$ |

*Figure 3.9 Incorrect instruction.*

replacing the subtract instruction with the add instruction. Obviously this program would not be difficult to debug, since it is short and simple. As you might expect, however, debugging can be a real challenge in situations where programs are long and complex. Obviously "bugs" are more likely to occur in such longer programs.

*The point.* The point of these last few pages is to illustrate, using a familiar example, that a computer does exactly what it is instructed to do and in a fixed order. If the instructions or the sequence is incorrect, the intended objective will not be met and it is up to the person who conceived, wrote, entered, and ran the program to find out why.

## SUMMARY

In this chapter you have been exposed to the rudiments of the discipline of software. Through the simple examples used, we have tried to put you in the place of the programmer, so that you might experience as much as possible the issues and concerns of those who write software, some of the problems that they face, and how those problems are solved. The treatment has been cursory but is in line with the overall objectives we are trying to achieve. Understanding the nature of software requires that you understand the technical concepts and some of the terminology.

# Objectives

At the conclusion of this chapter, you should be able to

1. discuss the concept of computer languages
2. discuss the advantages of higher level languages over machine language
3. define an operating system
4. discuss three higher level languages on an overview basis

# 4 Higher Level Language Concepts & Operating Systems

## INTRODUCTION

Chapter 3 demonstrated the creation and execution of a program. Each instruction in the program aided in the performance of a portion of the total task. The instructions used were written in machine language, a language that is directly "understood" by the computer's logic circuits and thus provides instructions directly executable by the CPU. As we saw in chapter 3, programming in machine language is somewhat cumbersome and complex, even to perform the simple task of adding two numbers.

Higher level languages allow programs to be written faster and easier than if the equivalent program was written entirely in machine language by the programmer. Because of the awareness surrounding the concerns and issues of creating a machine language program, you will really appreciate the programming ease and flexibility afforded by the use of higher level languages.

## THE LANGUAGE CONCEPT

The English language, one of the most complex in the world, is really nothing more than a culmination of many words. Each word has a specific meaning. Since we speak English and have mastered that language, we can communicate having already anticipated the results or response.

A higher level programming language can be compared in many ways to the text and usage of the English language. Higher level languages each have a set of words unique to that language. These words are generally more in line with plain English. Those words, or *keywords* as they are called, provide the means by which programs in that language can be written.

## HIGHER LEVEL LANGUAGES vs. MACHINE LANGUAGE

Higher level language words are not directly executable by a CPU's logic circuits. For example, take the word *print*, commonly found in a popular higher level language in use today, which will cause printing to take place on some output device such as a printer or a CRT. The word *print* does not correspond on a one-to-one basis with any machine language instruction in the machine language instruction set of a computer. In fact the computer's inherent capabilities do not recognize the letters *p-r-i-n-t* at all.

The phenomenon of higher level languages involves the use of a "translation mechanism" known as an *operating system*. The operating system takes higher level language vocabulary, such as the word *print*, and causes the results that we expect to occur through the creation and execution of a string of machine language instructions. Operating systems will be discussed in detail later, but to illustrate this point look at the case of a programmer who wants to write a program to print a message. The message is "hello." From the programmer's viewpoint, the task could be accomplished in two different ways:

- METHOD 1—by writing a machine language program to cause printing of the letters *h-e-l-l-o*
- METHOD 2—by writing a program in a higher level language to cause printing of the letters *h-e-l-l-o*

*Method 1: Machine Language*

Using Method 1 the programmer must do the following tasks (you will recognize some of these as we have already seen them done in chapter 3).

*Selection of proper instructions.* From the *menu*, or set, of machine language instructions documented in the computer's reference manual, the programmer must select the proper series of instructions that will cause the letters *h-e-l-l-o* to be sent to an output device such as a printer. The program containing these instructions should be documented on paper, with the instructions listed in the proper sequence. We know the importance of sequencing. As many as 40 to 50 machine language instructions might be required to do this.

*Selection of memory addresses for program storage.* We know that machine language programs must first be put into (stored in) the main memory prior to their execution. The programmer must therefore select the addresses in memory which will be used to store the program. Let's assume that 40 instructions are required to perform the task. Also assume that the programmer selects memory addresses 1–40 in which to store the program. Each instruction must be stored one at a time into the memory locations selected (see fig. 4.1.)

*Selection of memory addresses for data storage.* Just like the example in chapter 3, where data (two numbers to be added by the program)

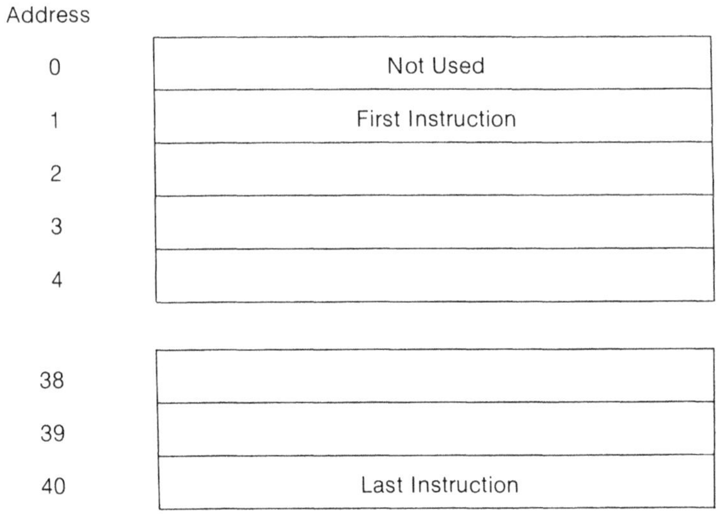


*Figure 4.1 Program entry.*

had to also be stored in memory, the programmer must, in this case, store in another area of memory a code for each letter to be printed (that is, *h-e-l-l-o*). There is a standard code in the realm of computer technology for letters, numbers, and characters such as #$%&'. The name for the code as a whole is ASCII, the American Standard Code for Information Interchange. As the name implies, the code is an industry standard. The logic circuits in printers (yes, they have them too) are designed such that when the code for the letter *h*, for example, is sent to the printer by the computer, the printer recognizes the code and prints the letter *h*. (See chap. 5 for more detail regarding ASCII code.)

So, the programmer selects five more memory locations. In each location he will store the ASCII code for a letter in the word *hello*. Assume that locations 50–54 are chosen. The programmer now inserts the code into the designated locations.

At this point, the program, stored in locations 1–40, and the data (letters), stored in locations 50–54, have all been entered. (See fig. 4.2.)

*Execution.* Now that the program is completely stored in memory it can be executed. The programmer does so and out on the printer comes the word *hello*. The programmer made no errors in program or data entry, therefore no debugging of the program was required.

*Summary—Method 1.* The programmer had to

- understand enough about the machine language instruction set for the computer he was working on to select and sequence the appropriate instructions to write the program
- understand enough about the system's memory so that he could intelligently select the memory area where the program could be stored
- understand enough about ASCII code so that he could correctly select the code for the letters *h-e-l-l-o*
- understand enough about the button-pushing mechanics of this computer so that he could enter the program and data into the computer's memory and start execution of the program

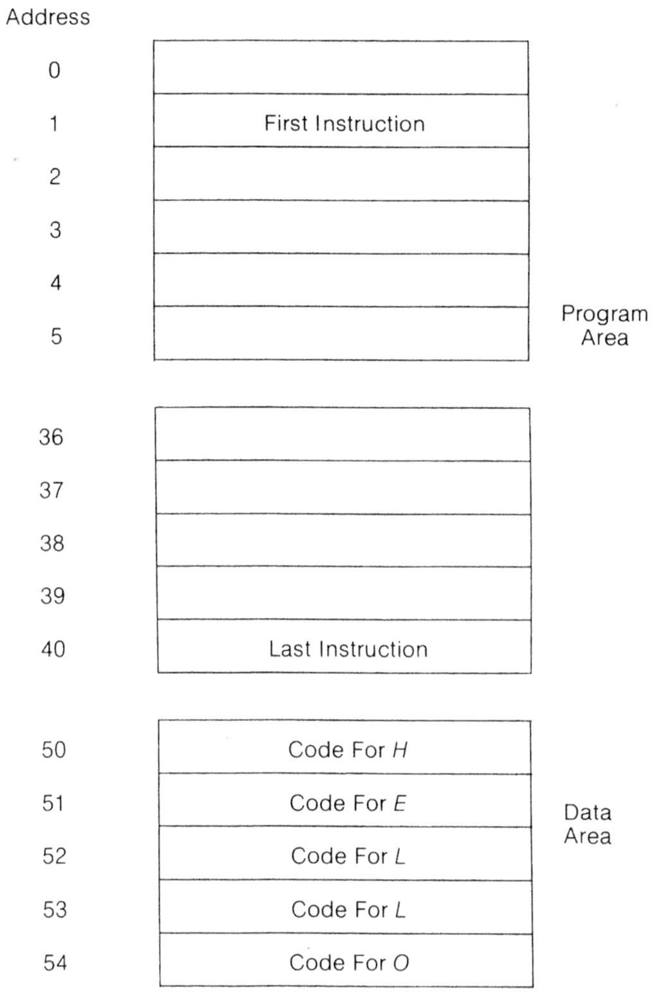


*Figure 4.2 Memory contents—program and data.*

Is this difficult and complex? For the experienced programmer with a knowledge of the computer in use, no. For you, on the other hand, yes. Yet it all must be done in order to achieve the seemingly simple task we set out to do.

*Method 2: Higher Level Language*

*Loading the translation mechanism (Operating System).* The programmer inserts a *disc* with the operating system stored on it into a disc unit. Discs are also discussed more in chapter 5, but for now think of a disc as simply a medium on which programs and data can be stored for future use.

The programmer then loads the operating system from the disc into the computer's main memory. This can be as simple as turning on the power switch of the computer. Most personal computers work this way.

*Statement entry.* The programmer enters, on a terminal with a keyboard, the statement:

PRINT "HELLO"

and out on the printer or CRT comes the word *hello*.

*Comparison of Method 1 and Method 2*

What is apparent is that Method 2 was a lot easier and achieved the same results as Method 1. The obvious differences are in what the programmer did *not* have to know or do. The programmer did not have to

- know anything about the machine language instructions that the computer uses
- write a machine language program
- worry about where (at what addresses) in memory the program would be stored
- store the program in memory
- know how to start the program running
- interface extensively with the button-pushing mechanics required to store and then execute the program
- know anything about bits, bytes, and so forth

Method 2 is obviously easier and requires a far less extensive knowledge of the intricacies of the machine itself. And that's the point! Use of a higher level language allows us to do far more in much less time than it would take to do the same thing using Method 1. It allows us to write programs, using a "higher level," easier-to-understand language, that are then automatically turned into the equivalent machine language programs required to achieve the desired results.

In fact Method 1 is truly being used in both circumstances described but in Method 2, unlike Method 1, we (as programmer) did not have to become involved. We did not have to be concerned about all the complex issues. All of the concerns were taken care of for us by that translation mechanism we called an operating system.

Now that we know what an operating system can do for us, we will examine it closer. Most of the details we could get into are not relevant to the objectives we are trying to achieve here, so this discussion will be a cursory one.

## OPERATING SYSTEMS

### *The System*

Based upon the example and comparison, you could easily conclude that an operating system is very complex in nature. The easiest way to define an operating system is to simply say that it is a very large machine language program. When loaded into computer memory, it occupies many thousands of locations. We've already referred to it as a translation mechanism. (Now see fig. 4.3.)

An operating system allows us to write programs in a higher level, easier-to-understand language. Computer logic circuits are only capable of executing a certain "repertoire" of machine language instructions. We learned in an earlier example that the computer does not directly "understand" the word *print* when entered, and yet when it was entered, printing would occur.

As shown in figure 4.3, the operating system is capable of accepting higher level language statements, or program input, that we type in, and "creating" the "lower level" machine language equivalent to do what we have input. The operating system examines our higher level language input, and implements and sequences correctly the corresponding machine language instructions.

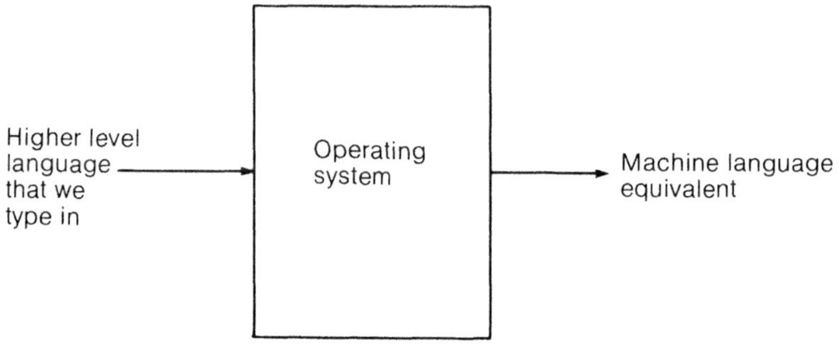


*Figure 4.3 "A translation mechanism."*

The true beauty of the operating system is that the user can be divorced of all of the concerns and cognitive skills required to program in machine language and instead, use a much simplified, higher level language to perform work on the computer. The user does not have to know anything at all about the operating system "nuts and bolts" or about the machine language instructions on the computer. The user does not even have to known that there is such a thing as machine language. All the user does is become familiar with a higher level language and then learn a few simple rules about how to type it in.

You should not assume that programming in a higher level language is a snap. If you go beyond the simple examples that we have discussed, you must first learn the language before you can write programs using it. This could take some time, but much less time than any alternative involving machine language programming.

*Languages and Operating Systems for Every Need*

The example we used earlier of a higher level language involved the use of the word *print*. *Print* is a word used in the BASIC (Beginner's All-purpose Symbolic Instruction Code) language. BASIC is one higher level language in common use today. There are others, over 100 others at least. Why so many? Why wouldn't one higher level language be sufficient for all programming needs? The answer to these questions really lies in the programmer's definition of the word sufficient.

Different programming environments require different strengths of the programming language used to solve the problems encountered in that environment. As a result of different needs, many different programming languages and operating systems have evolved. Three examples of higher level languages common today are

1. BASIC
2. FORTRAN
3. COBOL

*BASIC.* BASIC evolved as a result of the need to design a language generally comprehensive in its capability yet easy to learn. This would make it suitable for use by the beginner. BASIC is in common use today on many home computers. Institutions offer courses that can take a novice from the beginning to a rudimentary programming skill level in a few hours. Interestingly, many beginners know far less about the nuts and bolts of their own computers than you now know, yet they can write programs to do simple tasks after only a few hours of instruction. If you have a personal computer of your own, you may already be using BASIC.

*FORTRAN.* FORTRAN (FORmula TRANslation) has been in use for many years. It was developed to satisfy to a finer degree the particular concerns of scientists, mathemeticians, and engineers. They need a language which is mathematically powerful. FORTRAN is widely recognized as such a language. Mathematical modeling in research, for example, is facilitated using FORTRAN.

*COBOL.* Business has other special needs in a programming language. These needs encompass such tasks as calculating and printing the weekly payroll, keeping track of inventory, and so forth. COBOL (COmmon Business Oriented Language) is particularly suited to the needs of this sector and is commonly used as the primary language in corporate data processing departments.

## SUMMARY

This chapter has shown that one can program without becoming immersed in the technical and intricate nature and complex architecture of the computer itself. An operating system designed to support a higher level language is the key. We learn only the language. The operating system takes care of everything else required to turn our programs written in that language into a functioning reality.

# Objectives

At the conclusion of this chapter, you should be able to

1. define I/O devices commonly found on a microcomputer
2. discuss the function of a computer keyboard
3. discuss the use of a CRT and its capability
4. discuss printer concepts including types of printers commonly available, transmission modes, and communication protocols
5. discuss the printer interface and how to know whether or not one is required
6. discuss buffered printing devices and cite their advantages
7. discuss disc devices with regard to their purpose, and generally how they function

# 5 Input/Output Devices

## INTRODUCTION

What do we know about the input and output devices that computers use? We know that computers use input devices such as keyboards so that we can input higher level language statements, for example, for use in programs. *Output* devices are used to convey output, such as messages ("hello"), reports, and displays of data, all for our use. But what are these devices? What kinds of generalized I/O (input/output) devices exist? That is the subject of this chapter.

## KEYBOARDS

You probably already have a familiarity with the input device called a keyboard. It is an array of keys with a character printed on each. Stroking a key will cause the associated character to be sent to the computer. Whether the character sent is "recognized" by the computer or acted upon in any way, depends on whether software currently running in the computer is "looking" for input from the keyboard. Sometimes software does not require, or "look for," keyboard input. Pressing a key on the keyboard at such times may therefore produce no visible effects. An example of keyboard interaction is shown in figures 5.1–5.3.

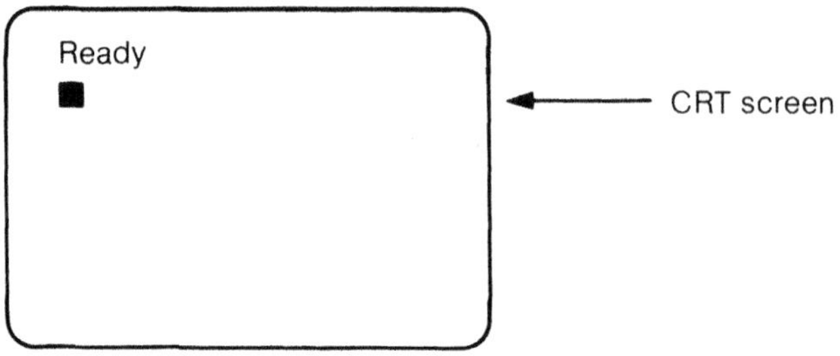


*Figure 5.1 CRT screen showing* ready *and cursor.*

In figure 5.1, the user has inserted a BASIC operating system disc into the disc unit, turned the computer power on, and loaded the BASIC operating system into the computer's main memory with the intent of entering a program written in BASIC. The visual prompt *ready* shown on the CRT screen indicates that the operating system has been loaded, is now running, and is ready to accept keyboard input. The black square below the *ready* prompt is known as the *cursor*. It indicates where on the screen a character will be next displayed when it is sent to the CRT.

The software currently running in the computer (i.e., the BASIC operating system) is waiting for the user, or programmer in this case, to enter a statement on the keyboard. In figure 5.2 the user has entered *10 print "hello"* on the keyboard. As each character is entered, the cursor moves to the position on the screen where the next character entered will be displayed. It should be noted that as each character is entered from the

*Figure 5.2 CRT screen showing entry of statement and cursor position.*

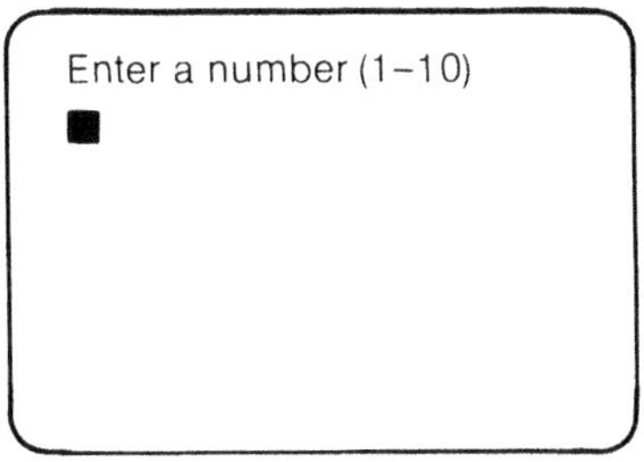


*Figure 5.3 CRT screen showing prompt asking for data input.*

keyboard, the computer (software) is designed to send, or echo, as it is called, that character to the CRT which causes the corresponding display to occur. This acts as visual verification that the character depressed on the keyboard has in fact been input to the computer (recognized) and stored.

The *10* entered first in our example is known as a statement number. Each statement in a BASIC program has a statement number. The statement number defines where that statement will be sequenced and executed later in relation to all other statements in the program. Statement number 5, for example, would precede statement number 10, which would precede statement number 20. When the program was executed, statement number 5 would be executed before statement number 10, and so forth.

In figure 5.3 a program is running that asks the user to enter data (i.e., a number from 1 to 10). Such data is entered using the keyboard.

## CRTs

CRTs (Cathode Ray Tubes) are used to generate displays for the user's use. Specific examples of how a CRT is typically used were just shown. Many CRTs are strictly used as display terminals and are therefore only capable of output, that is, data is sent only from the computer to the CRT. Other CRTs, however, including many used in computer-based training applications, have an input capability (ability to input characters) as well as an output capability. These are discussed more fully in chapter 9. Specific features to look for in a CRT are also discussed there.

## PRINTERS

Printers are used to print hard copy (paper) documents and as such are widely implemented in a variety of applications. There are several basic concepts about printers in particular that you should be aware of, as your future evaluation of units may depend on it.

### *Communication Protocol*

Since the text printed by a printer is sent to it by a computer, it is essential that the computer and printer be capable of communicating successfully. That is, if the computer sends the letter *c* to the printer, it is crucial for the printer to recognize it and print it correctly. This sounds like a simple concept on the surface but in fact is not.

A successful communication link between a computer and a printer is made more difficult because many computer systems feature I/O devices from a mixed bag of manufacturers. For example, an ATARI 800 computer system may be connected to a printer made by EPSON. Chiefly, for this reason, it was recognized years ago that an industry standard communication protocol needed to be conceived and adopted by all industry manufacturers to allow effective communication between devices of different make.

The American Standard Code for Information Interchange (ASCII) is one such standard although it is not the only one. Using an 8-bit format, any letter, number, or special character, such as !"#$%&, can be sent from device *A* and correctly interpreted by device *B*. Bit combinations assigned to a character are unique to that character. Therefore each character has its own integrity making it impossible for properly functioning printer devices to "confuse" one character with another. One example of an ASCII character for the letter *A* is shown in figure 5.4. Note that bits 8, 7, and 1 are high (1) while all the rest are low (0). When these bits, in this combination and sequence, are

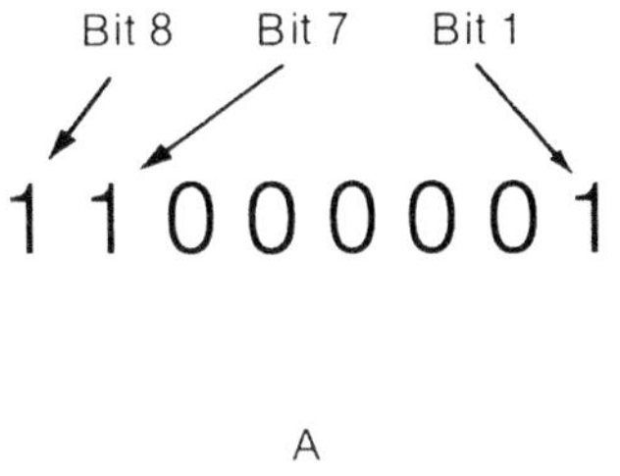


*Figure 5.4 An* A *in ASCII.*

sent to a printer, the printer should print the letter *A* in upper case. Provisions in ASCII also allow the transmission of lower case letters.

### *Transmission Mode*

We know that bits are transmitted on wires (or buses) as high or low voltages. ASCII bits are no different and are transmitted over a cable, containing wires, from a computer to a printer. This is how all characters are sent to a printer.

When you think about it though, there are really two ways that an eight-bit ASCII character could be sent to a printer. One way is to transmit eight bits over eight wires, all at the same time, such as shown in figure 5.5.

A second way is to send the same eight bits on one wire, but sending them one at a time. This would require eight time periods to send all eight bits, one after another. Allow that we cannot send two or more bits on the same wire at the same time.

The first method, demonstrated in figure 5.5, where eight wires are used and all eight bits are sent at the same time, is called *parallel transmission.* You can now see how the term was coined. The advantage of parallel transmission of characters is that the whole character (all eight bits) can be transmitted at once during the same time period.

The second method, where one wire was used to send all eight bits, one at a time, is called *serial transmission.* An advantage that serial transmission enjoys is that less hardware (wires) is required to send eight-bit characters since only one bit may be sent at a time; thus, only one wire is required.

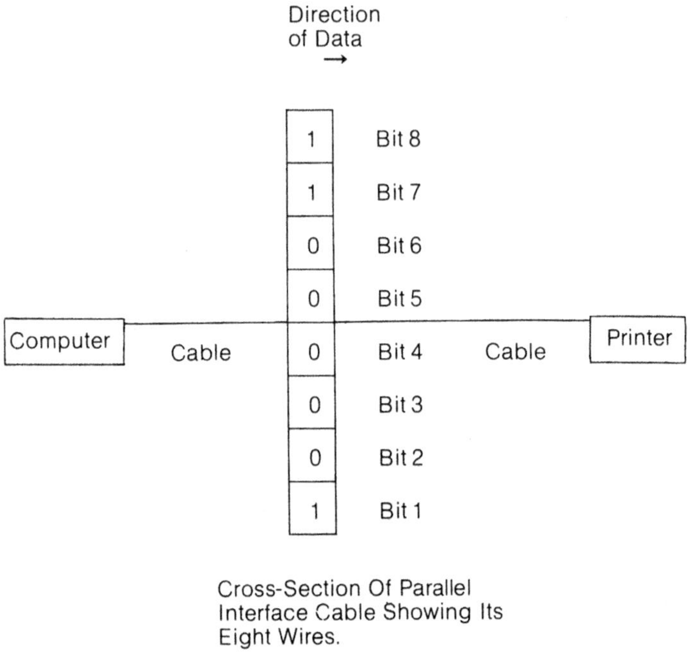


*Figure 5.5 Parallel transmission.*

Some printers require parallel input and will not work with serial input. Many computers, on the other hand, send characters in serial format. This presents an obvious problem. To solve the problem, a specific set of logic circuits known as an *interface* is placed electrically between the computer and the printer. This is shown in figure 5.6. This interface may contain a special logic circuit known as a SIPO (see-poe), which stands for Serial In—Parallel Out. A SIPO changes serial input to parallel output.

If connected as shown in figure 5.6, the serial format output by the computer is changed to parallel format before arriving at the printer. With the use of the interface, different transmission modes can be made compatible, allowing effective communication between computer and printer.

You should know, however, that an interface is not always required because

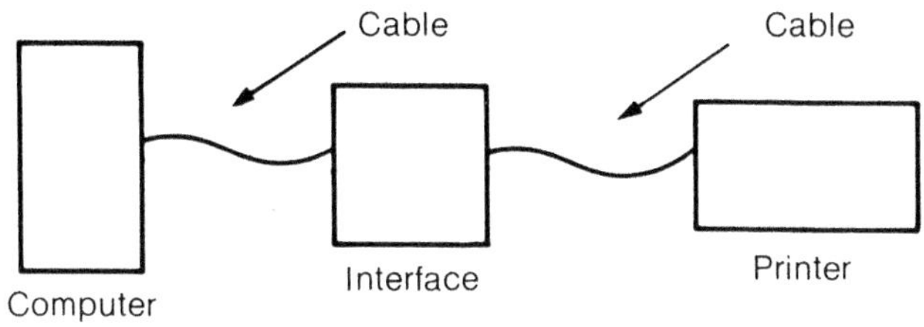


*Figure 5.6 Interface connection.*

- some computers can transmit in parallel or serial fashion
- some printers have the ability to receive either parallel or serial data.

Computers that can transmit in either mode have more than one cable connection, or *port*, for a printer. The connection you use to connect your printer cable depends on what your printer input requirements are. Printers that can receive either parallel or serial data have an interface function within the printer housing. Usually a switch determines whether or not the interface function is actually used.

*Print Format*

Two common print formats are available on printers for microcomputers. They are

- dot matrix
- letter quality

*Dot matrix.* Dot matrix printers use a matrix of dots to form characters. Dots are made using a group of wires housed in the print head. Each wire is individually actuated by a solenoid. Actuating solenoids in various combinations causes corresponding wires to make a dot impression on the paper, through a ribbon much like a typewriter ribbon.

Various resolutions of dot matrix printers are available. Two common configurations are $7 \times 9$ and $9 \times 9$. The $7 \times 9$ configuration has the capability to print 7 dots horizontally and 9 dots vertically in each character position. The 9 × 9 configuration can print 9 dots horizontally and vertically in each character position. The higher density dot matrix configurations produce a character of higher resolution on paper.

Dot matrix printers are usually faster in terms of the number of characters per second that they print when compared to letter quality printers.

*Letter quality.* The letter quality printer does not use a dot matrix concept at all but rather makes actual letter impressions much like a typewriter. One type of letter quality printer uses what is known as a "daisy wheel." The daisy wheel itself is a circular device with many "fingers," each of which has a character on it. The wheel rotates quickly to bring each letter in line with a striker. The striker then strikes the finger with that letter on it and an impression is made on the paper, again using a ribbon.

Daisy wheel printers are characteristically slow because of the time it takes to constantly rotate the wheel between letter impressions. The advantage they have is that the resulting text printed is of "letter quality" and thus of higher quality than a dot matrix printer.

Letter quality printers are usually much more expensive than a dot matrix printer. One item to consider, therefore, is whether or not the extra expense is a requirement in your application.

### *Buffered Printers*

There is much talk surrounding the concept of printers with and without buffers. A buffer is a small memory inside a printer. The purpose of the buffer is to allow the printer, which is always much slower than the computer, to accept and store characters to be printed as fast as the computer can send them. The printer then prints from the contents of the buffer, while the computer goes on to some other task.

The advantage of having a printer with a buffer is that the user does not have to wait until a document is completely printed before returning to work on the computer. In situations where the computer is outputting to a printer without a buffer, the computer does not usually allow other work to be done by the user. Thus the user has to wait until the document has been printed. Simply stated, output from the computer is occurring roughly at the rate at which the printer can print. When outputting to a buffered printer, however, the computer can fill the buffer very quickly, as fast as it can output, without regard to the speed at which the printer actually prints the information. This means that the computer can finish sending data to the printer much faster than in the case of an unbuffered printer and be ready to allow the user to work on other documents, while the printer prints the document just output. Needless to say, buffered printers are generally more expensive than unbuffered printers.

A practical consideration in selecting a buffered versus an unbuffered printer is the print speed of the printer. For example, daisy wheel printers are generally very slow (10 characters per second). If you plan on working with multiple documents and have a very slow printer, it may be advisable to buy a buffered printer to save you time. While one document is being printed, you can be working on the next.

## DISCS

Discs and disc units are perhaps the cornerstone, in many minds, of the computer system of today. In short, discs are used to save programs and data in a more permanent form than is possible if those same programs and data were to be left in the computer's memory.

Programs and data left in computer memory are often subject to corruption. If the power to the computer system fails, for example, programs and data in storage at the time of the failure may be "wiped out" due to the nature and technology of the memory that stores them. Computer memory of the integrated circuit variety requires the application of power in order that it may "remember" data. The lack of power causes "amnesia" in these types of memory and may cause destruction of hours of work on the part of the programmer.

To eliminate the possibility that this might occur, programmers rely on discs to store programs and data generated initially in the computer. By "saving" a program on a disc, the programmer is transferring all program-relevant information to that disc. He can then take the disc with him and store it in a safe location.

*The Disc Defined*

A disc is a magnetic platter, housed in some form of case. Some cases are flexible. The associated discs are known as *flexible discs* or *floppy discs*. Other discs are encased in a hard disc pack, which is not flexible. Discs can store a varying number of bytes of equivalent computer memory. Usually flexible discs can store much less than hard discs or disc packs. Nonetheless, they all serve the same purpose, which is to provide a safe storage medium for programs and data created initially in the computer's memory.

*The Disc Unit or Drive*

In order that data or programs may be initially written on a disc, a disc drive is required. A disc drive is an assembly of electronic logic circuits and electromechanical components that serve the purpose of writing data on discs. Included in the disc drive is a *head* or heads that are applied to the disc surface or *fly* just above the disc surface.

The job of the heads is to write data on the disc or read data from the disc. The disc drive will write data or programs on the disc when a command is initiated in the computer, by the user, to *save* a program or data. The disc drive will read data or programs from the disc when a command is initiated by the user to *load* data or programs from the disc into the computer's memory. All such discs are considered auxiliary storage devices.

Reiterating, the advantage of the disc is that once programs or data have been saved, the disc can be removed to a safe location. Since data is magnetically stored on a disc, no ongoing power source is required to maintain the integrity of data stored. This is contrasted to the computer's main memory, which requires that power be constantly applied in order to maintain data integrity.

It should be noted that there are several different "technologies" of main memory in computers. Some do not require the application of power to maintain data. Core memory, which uses very small ferrite (iron) rings to store bits, is an example. Ironically, the newer technology memories consisting of integrated circuits do require constant application of power in order to maintain data stored.

*Disc Organization*

When viewing the intricacies of how data and programs are stored on a disc, it becomes apparent that the computer and the disc must maintain some level of organization so that the computer can later "find" data that it has stored on the disc. For this reason, discs are organized into *tracks* and *sectors*, which serve to divide the disc up into parts that are identifiable by both the computer and the disc drive logic circuits. Tracks are defined on a disc as circles of ever decreasing circumference. Sectors are visualized as radii emanating from the center of the disc, to divide the tracks into equal portions, similar to dividing a pie into equal pieces.

A typical disc and its organization is shown in figure 5.7. The number of tracks and sectors on a disc vary from disc to disc.

A new disc must be formatted before it can be used to store data or programs. The process of formatting establishes the partitions on the disc that are required for a particular computer system to successfully store data or programs on that disc. Formatting is done through the execution of a special program

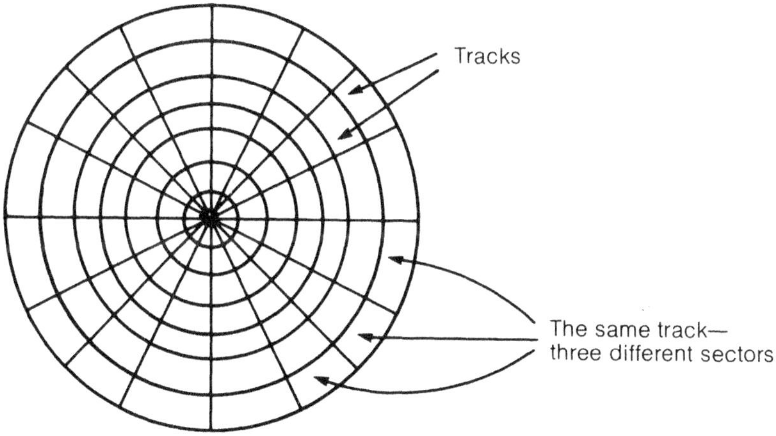


*Figure 5.7 Disc organization.*

that does nothing more than "divide the pie." You should understand that discs formatted on one system type may not be usable by another system type, since formatting procedures and specifics change from system to system.

### *Use of a Disc*

Preparing a system to use a disc is simply a matter of attaching the disc unit to the computer. This is done using an I/O cable to connect the two. Next, the programmer must format the discs to be used for program and data storage by loading and executing a formatting routine (small program) that formats the discs once inserted into the disc drive.

Having formatted a disc, the programmer can now create software in the computer's memory, and later save it out to the disc. Once programs and data have been saved to the disc, the computer system can be turned off without fear that the data has been lost (short of typing it all in again). This scenario is the normal expectation of users of personal computers who tend to want to turn their computers on and off at will.

## SUMMARY

This chapter has examined some of the more typical I/O devices found on a generic microcomputer system. The direct impact on you can be recognized through examination of the following questions:

- Do I need a keyboard?
- Do I need a CRT? If so, does the CRT have to serve an output function (display) only or do I also need an input capability (e.g., keyboard)?
- If I need a printer, do I need a letter quality printer or will a dot matrix unit serve the purpose for less money?
- What is the transmission mode of my computer (serial or parallel)? How does that impact my printer selection? Do I need an interface?
- Do I need a buffered printer or is an unbuffered version suitable?
- Do I need a disc device?

# □ Objectives

At the end of this chapter, you should be able to

1. define a microprocessor
2. define a microcomputer system
3. define and discuss the technologies that have led to the advancement of the microcomputer system
4. define a typical microcomputer system and its components

# 6 Microcomputers and Microprocessors

## INTRODUCTION

It seems as though we have immersed ourselves in some detail already regarding what a computer is. By many standards, however, we have barely scratched the surface. But hopefully you have come to grips with the computer as a technology. We now know the computer for what it is and as it really exists. We also know a bit of what is required of the people who make it work its "wonders." If nothing else, we now know that people are required.

Recently, in the last ten years or so, a new technology of computer has come into its own; a computer that possesses all of the characteristics we have discussed yet is smaller and far less expensive than those "beasts" we always seem to envision. Is it tamer? Technologically no, which is significant, but it seems to be kinder in appearance to many people I have talked to. It is the *microcomputer*.

## WHAT IS A MICROCOMPUTER?

In the early days of computing (the 1950s), computers were large. Chapter 1 described the ENIAC, the first computer. It was huge and took up much floor space. It was a marvel of its era.

Today technology has progressed to such a point as to make computers much smaller and less expensive.

Setting the technology aside for a moment, a microcomputer is simply a small computer. Everything it contains is really a miniaturization of the counterparts of larger machines. It can be set on a desktop with room left over. It can be powered from a single wall plug in your home. It is much less expensive than anything we could previously imagine. For all of these reasons, it has become a viable personal entity and resides in many homes and businesses today.

One misconception to be dispelled, then, is the notion that the microcomputer is a new and completely different computer than the larger computers. It is not. It is the product of much research and the discovery of new techniques to make smaller what was larger before. Engineering science is to be congratulated for its many and significant achievements that have led to the realization of the computer that is revolutionizing our perception and use.

In short, a microcomputer is a smaller and less expensive computer. Let's discuss briefly these attributes.

*Size and Cost*

We know that the fundamental building blocks of a computer system's hardware are

1. the CPU or processor
2. the memory
3. input devices
4. output devices

*Size.* Through a series of technological advances, the physical size of a CPU has been reduced from occupying a few hundred square feet to just a few square inches. The actual size of a typical microprocessor (i.e., the CPU portion of a microcomputer) is shown in figure 6.1.

This miniaturization has been made possible through the development of the technology of circuit integration where the functions of many thousands of components are now reduced in size but not in functionality. The latest techniques of miniaturization employed in this field are known as LSI (large scale integration) and VLSI (very large scale integration). This has

*Figure 6.1 Microprocessor chip, actual size.*

simply reduced the space required to house those complex functions, which in many ways have existed for years.

Technology has also generated a way in which these integrated circuits (ICs) can be produced in volume. The cost, therefore, has been dramatically reduced with no corresponding loss in function or capability. In fact, personal computers, as small as they are now, will be still smaller in the years to come. Without much doubt computers will be commonly available as hand-held devices within the next two to four years.

*Cost.* Due to the evolution of the science of miniaturization, as well as our ability to mass produce microprocessors, the cost of manufacturing has been reduced to a few dollars per unit. The price paid for a personal computer today, as low as it seems, is not even based on the cost of the components as much as on the labor involved to assemble them into a working system.

## MEMORY TECHNOLOGY

Memory used to be big too. Memory technologies developed in the seventies only allowed us to put 16K bytes on a printed circuit board "larger than a breadbox." Now 64K of memory can reside on a single integrated circuit chip the size of that shown in figure 6.1. Incredible? Yes! Memory costs as high as a few dollars per byte years ago are now reduced to a few dollars per chip (a chip houses thousands of bytes of memory).

Memory technology continues to advance with the emphasis on size reduction. A leader in the field has recently introduced a single integrated circuit chip that houses 128K of memory.

## A TYPICAL MICROCOMPUTER

A typical microcomputer installation is shown in figure 6.2. One small desktop unit usually houses the microprocessor, or CPU, as well as a keyboard for program and data entry. Cables connect the microprocessor unit to

- a disc unit that houses electronic circuitry for reading data and programs from a disc and writing programs and data on a disc. The disc is inserted into the disc unit, or drive as it is commonly referred to, when reading or writing is required. (See chap. 5 for more detail.)
- a CRT (cathode ray tube) used for viewing displays generated by the computer
- a printer for generating hard copy, such as reports

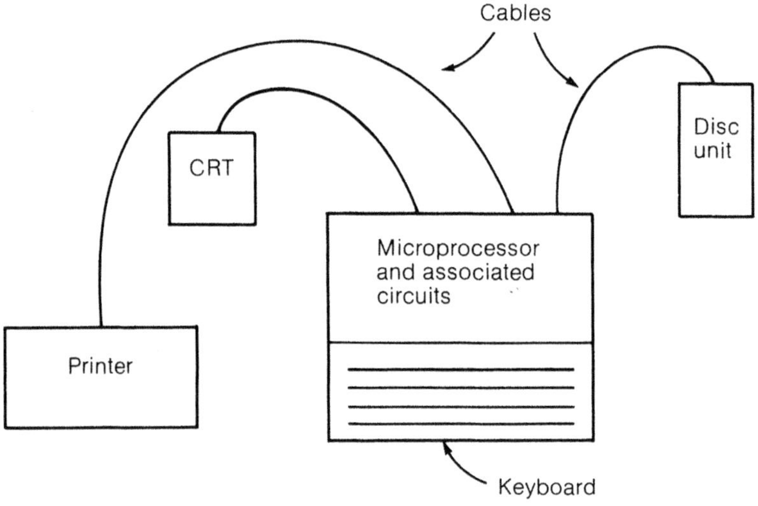


*Figure 6.2 Typical microcomputer system.*

## SUMMARY

Microcomputers are popular. Why?

- They are small and yet accommodate operating systems that allow a friendly programming capability.
- The advancements in the technology required to produce them has brought the cost of microcomputers down and within the reach of many of us.
- Their small size further adds to the feasibility of implementation in many work and pleasure environments.
- Advances in memory technology have produced memory of respectable size in a small package the size of an integrated circuit. This makes the use of the microcomputer more attractive because given the memory capabilities, the applications are many and varied.

# Objectives

There are no specific learning objectives for this chapter. Chapter 7's goal is to review the contents and material covered in detail in chapters 2–6. It will act as a handy and concise reference source for future use.

# 7 System Summary and Review

## REVIEW OF CHAPTER 2—HARDWARE

How did chapter 2 define hardware? What comprises hardware? What are the basic hardware-related issues?

DC voltage, not unlike that found in a battery-operated toy or appliance, acts as the medium of intelligence in computers. Computer hardware is incapable of thought as we know it, but through logic circuits it can make simple decisions. Logic circuits are combined by the thousands to enable computers to simply process instructions, one at a time, very rapidly. These instructions are input to the computer's memory component, and later extracted from memory and executed by the central processing unit.

Instructions and data are manifested physically in the computer as a unique combination of voltages, some high and some low. The interpretation of this unique combination of voltages is done by logic circuits. Voltages, not thoughts, make the logic circuits of a computer work. Although only two voltages are used, their combination in a positional framework allows us to represent many things. Examples included the representation of eight colors.

The memory component of the computer was introduced next, with its structure of many locations. Each location in memory can contain an instruction or byte of data.

We learned to count using the binary numbering system where only two valid numbers exist, 0 and 1. There is a positional framework in the binary numbering system just as in the decimal system. Adding positions allows us to count higher and higher, just as in the decimal system. The two voltages used in computer logic circuits are well suited to the two valid numbers in the binary numbering system and are equated.

Memory addresses, unique to each location in memory, are assigned to memory locations. This addressing scheme allows each memory location, and there may be thousands of locations, to assume a unique identity we can refer to.

Memory locations are comprised of a certain number of bits. Each bit can contain a portion of an instruction or a number, which is referred to as data. The number of bits per memory location will always be the same in a given memory. Different memories can have any number of bits per location, or byte. Typical memories today have either 8 or 16 bits per byte.

Large memories, having thousands of memory locations, are really no different than smaller memories except that to address all of the locations in a larger memory, a larger address scheme (more bits) is required. Do not confuse the address itself with the contents of the memory location identified by an address. For example, at address 00000, 11111111 may be stored. Since there are 5 bits in the memory address here, the memory size, or length, is 32 bytes ($11111_2 \equiv 31_{10}$ and address 00000 makes 32 addresses.). The memory width, or the size of each memory location, is 8 bits.

A central processing unit (or CPU) is defined as that portion of the computer system that executes instructions stored in memory, and manipulates data stored in memory. The CPU uses many logic circuits, including *AND* gates, *OR* gates, and inverters. By combining the functions of each of these types of gates in specific configurations, a unique combination of inputs will cause an output or set of outputs to be generated.

In addition to the memory and CPU components of a computer system, input and output devices allow the user to communicate with and receive output from the computer.

Buses link all major components of a computer system. These buses are nothing more than a group of wires that allow transmission of instructions and data (voltage combinations) to and from the the different major components of the system. The wires of the buses carry the DC voltages that we know are the representation of all intelligence within the system.

## REVIEW OF CHAPTER 3—SOFTWARE

In chapter 3 the hardware principles discussed in chapter 2 were used as a foundation to explore the creation and implementation of software.

A problem or a specific need always precedes the creation of software. Once the need is recognized and defined, a programmer can devise a program to satisfy the need. Full implementation involves not only writing the program, but entering it into the computer's memory, and then running it.

Not all computer programs run satisfactorily the first time due to the existence of "bugs" in the program. Bugs may arise as a result of the incorrect sequencing of instructions or the insertion of an incorrect instruction. Only the programmer, through examination, can resolve bugs in software. This can be a time-consuming and somewhat frustrating task as well, particularly if the program being "debugged" is long and complicated.

Program writing in machine language, or using the instruction set (menu) that the computer can directly execute, requires the programmer to first document the steps required to solve the problem or satisfy the need. Instructions from the reference manual can then be selected and sequenced to accomplish each defined step. The programmer must next select the area of memory (i.e., which locations) to be used to store the program. The program is then entered into memory. The entering process is somewhat mechanistic and varies from computer to computer.

Once the program has been entered, along with all associated data to be used by the program, it can then be executed. The program counter, which is the access mechanism to any and all memory locations, is oriented to the memory location containing the first instruction in the program. The computer is then started, and hopefully the program will run to completion without any problems. If problems are encountered, the programmer must then debug the program as indicated above.

## REVIEW OF CHAPTER 4—HIGHER LEVEL LANGUAGES

The concept of higher level languages can be likened to the concept of our own English language. We speak in English using words, which have predefined meanings, in order to communicate. Each higher level language, and there are many, has a set of keywords, each of which has a defined meaning and which will invoke certain defined actions understood by the programmer. The keyword *print* for example, as used in BASIC, will invoke the action of printing text or data on a printer or a CRT.

Use of higher level languages is a much more desirable way to write programs because the language is easier to understand and because many of the complexities of programming in machine language now become of no concern to the programmer. For example, the programmer programming in machine language had to directly interact with memory in the selection of memory addresses in which to store and enter the program. The same job or task can be accomplished much more readily by using a higher level language in that the inexperienced programmer need only learn the language and some mechanics about entering programs from a terminal.

Programs written in a higher level language are eventually transformed to the equivalent machine language and executed by the computer, but the programmer does not have to write the machine language or worry about where in memory (i.e., what locations) the program will be stored. These concerns are taken care of by the operating system under which the higher level language is operating.

When writing programs in a higher level language, the programmer need only load the operating system and begin programming, entering program statements in that language. When a higher level language program has been entered through the programming terminal, it can be executed without any direct hardware interaction. At run time, or execution time, the operating system translates the higher level language program statements into appropriate machine language automatically.

Many different higher level languages such as BASIC, FORTRAN, and COBOL exist to satisfy the particular programming needs of various disciplines and environments, such as those found in personal computer, scientific, and business environments.

## REVIEW OF CHAPTER 5—INPUT/OUTPUT DEVICES

Input/output devices are required in order to allow the user to enter program statements or data into the computer, and receive the results of program execution such as printouts on a printer or displays on a CRT.

Keyboards are used for program and data entry. In many cases, input from a keyboard is echoed to a CRT so as to allow visual verification of entry.

CRTs (cathode ray tubes) constitute the main display device in a computer system. They are used to generate displays by which the user observes and interprets the data/text presented and then perhaps responds by entering more data on a keyboard. Some CRTs with plasma or "touch" screens also have a capability for input. Such special devices are explained in more depth in chapter 9.

Printers are used to generate hard copy documents. Many printers use industry standard communication protocol, such as ASCII. This allows printers and computers of different manufacture to be connected together and communicate successfully.

Interfaces are sometimes required and connected between the computer and the printer so as to satisfy the differences between data communication expectations. While some computers send data to a printer in serial fashion, one bit of the character at a time, some printers require their input to be in parallel form, that is, all character bits being received at the same time on different wires. Interfaces are used to convert serial to parallel so as to close the "communication gap" that would otherwise be created and present a problem.

Some printers use a dot matrix print style, while others feature letter quality printing. Dot matrix printers have an array of wires, each actuated by a solenoid. Various combinations of actuated solenoids cause their corresponding wires to make a dot impression on paper, forming letters, numbers, and so forth.

Letter quality printers feature whole characters, mounted on a daisy wheel or other similar device, and are struck by a striker, making an impression through a ribbon on paper. The quality of the characters produced is better than dot matrix since the characters are continuous in nature, such as on a typewriter, and not formed by an array of dots. Letter quality printers are generally slower than dot matrix printers due to the manner in which they print. They also tend to be more expensive than dot matrix printers.

Some printers have buffers, or small memories, which facilitate faster transfer of data from a computer to the printer. Computers can send data to a printer at their own rate, rather than at the printer's much slower rate. This frees up the computer for other work, while the printer prints out the preceding document.

Disc units, or drives, are used to save programs or data on discs. They are also used to read, or load, programs or data from the disc back into the computer's memory, so that the programs can be executed. Discs come in several varieties and types. The one most commonly known to the general public is the floppy or flexible disc.

The floppy disc is a circular magnetic entity that looks something like a record. It is contained within a jacket, or envelope, which is usually black in color. When the disc is inserted in the disc unit, it begins to spin. Recording heads, located in the disc unit, are applied to the surface of the disc. Programs and data can then be read from the disc or written onto the disc.

Discs are organized into partitions known as tracks and sectors. This organization is needed to allow the disc unit and computer to "know" where specific information is located on the disc. In this way, data or programs may be located by the computer with the help of the logic circuits in the disc unit.

## REVIEW OF CHAPTER 6—MICROCOMPUTERS AND MICROPROCESSORS

A microcomputer is simply a small computer. It has been made possible in recent years as the result of advancing technology and the discovery that logic circuits can be compacted without a decline in function. This technology of microminiaturization has produced entire CPUs, called microprocessors, that are but a few square inches in size. It has also paved the way toward squeezing much more memory in much less space. A basic microcomputer system, made up of a microprocessor, memory, and input/

output circuits, can now all reside on one relatively small printed circuit board.

Due to the advancing technology described, not only has the size of such systems been dramatically reduced but so has the cost, bringing the computer well within the range of small businesses and consumers alike. A typical microcomputer system contains a microprocessor and memory, a CRT, a printer, and a disc unit. The disc unit uses discs on which programs and data, initially created in the computer, can be saved for future use.

# PART II

# THE USE OF COMPUTERS IN TRAINING AND DEVELOPMENT

# Objectives

At the conclusion of chapter 8, you should be able to

1. define the most commonly used terms associated with using the computer in a training environment
2. identify the most common uses of the computer in a training environment
3. identify the types of software programs available for use in training
4. identify the basic hardware necessary for implementing the computer in a training environment

# 8 Concepts of Computer-Based Instruction

## INTRODUCTION

The first part of this text dealt with the basic concepts surrounding the operation of the digital computer. It was intended to acquaint you with the computer from an operational perspective. You have been familiarized with much of the terminology surrounding this exciting science. Now that you have this background, we will take a look at how to put it to use in the training and development environment. The primary goal of this chapter is to familiarize you with the terms and general concepts that you will encounter in more detail in later chapters.

## TERMS FOR TRAINING INVOLVING COMPUTERS

Acronyms are widely used in today's high technology areas. This holds true for the terms associated with the use of the computer in training and development. Terms such as CBT (computer-based training), CBI (computer-based instruction), CAI (computer-aided instruction), and CMI (computer-managed instruction) are popular. Their meanings and implications often overlap depending upon who you talk to. In this chapter the terms that are most often used in the training industry will be defined, recognizing that "pockets of deviation" do exist.

All of the acronyms associated with the use of the computer in training can be grouped into two major categories: (1) computer-aided instruction and (2) computer-managed instruction. The first term, *computer-aided instruction*, or CAI, is defined as the use of the computer strictly as an instructional medium. This means that the computer is used as the delivery tool and is usually applied as a stand-alone training mechanism where a human instructor is not the primary delivery agent.

The second application, *computer-managed instruction*, or CMI, is defined as the use of the computer to manage or support the training process. In this case the computer may or may not be the delivery agent. In fact the student may never see the computer.

Both applications are collectively referred to as *computer-based training* or CBT. Computer-based training is a general term used to refer to the implementation of the computer in delivery or support or both. Figure 8.1 shows the interrelationships of CAI, CMI, and CBT. As you can see, computer-based training encompasses both major uses of the computer in training and development.

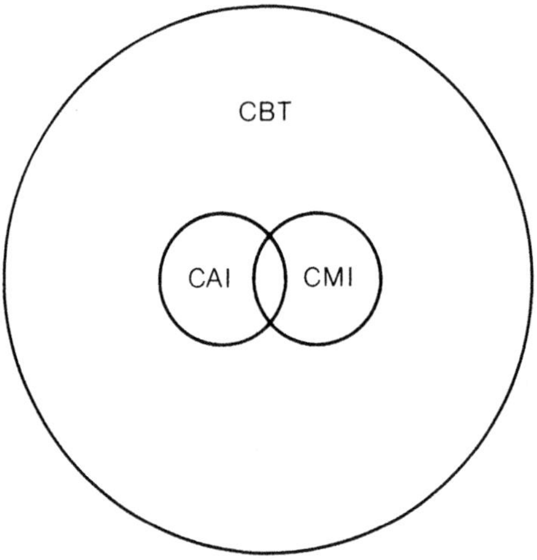


*Figure 8.1 Interrelationships.*

## COMPUTER-AIDED INSTRUCTION—CAI

To date, CAI has been the most visible application of the computer in training. It is characterized by the total involvement of the student with the computer. There is normally little or no outside intervention during a segment of learning when CAI is employed, since the computer is the delivery agent.

CAI can be subdivided into six unique applications:

1. drill and practice
2. tutorial
3. simulation
4. problem solving
5. informational
6. instructional gaming

### *Drill and Practice*

Drill and practice involves the delivery of small portions of information the student must learn. Learning lists of terms or definitions and the matching of generic terms with specific terms are examples that fall into this category.

Drill and practice are based on the theories of classic behaviorism. A stimulus, such as a term, is presented to the student who is then asked to indicate the correct response. The design of the supporting software has some flexibility so that the stimulus can be presented at random or according to a predesigned scheme.

Drill and practice exercises are useful for learning basic terms or for presenting simple procedures for memorization. Drill and practice methodologies are generally not used for teaching complex procedures or methods of operation. An example of a simple procedure that could be taught using the drill and practice method is shown in Figure 8.2.

### *Tutorial*

A tutorial program is modeled after classic programmed instructional methods. That is, some information is presented in module format, then the student is questioned for understanding. The information is presented in units called *frames*. If done creatively, these programs can be extremely useful in the instructional setting.

Tutorial programs can be used to guide a student in learning procedures of a more complex nature, methods of operation, or

EXAMPLE

List The Steps For Loading A Camera.

1. ________ Insert Film.

2. ________ Open Back.

3. ________ Close Back.

4. ________ Wind to First Exposure.

5. ________ Attach Leader.

*Figure 8.2 Drill and practice.*

even specifications on products. Information can be presented in a straight line fashion, known as linear order, or by branching through a series or topics or remediation, depending upon the correctness of the student's answers. An example of the use of tutorials is shown in figure 8.3.

*Simulation*

A CAI simulation is a program that models or recreates a real life situation. It is employed whenever the real experience would represent a dangerous learning environment, is not cost effective, or is unavailable to the students for learning.

Three general types of simulation can be identified. They are

1. performance simulation
2. systems modeling
3. experiential simulation

EXAMPLE

Light-sensitive materials, when exposed to light, will change. Photographic film and printing papers are examples of light-sensitive materials. When exposed to_____, these light-sensitive materials will_____.

(ans.—light, change)

*Figure 8.3 Tutorial.*

*Performance simulation.* Performance simulation teaches the student a particular skill. One of the best-known examples of this type of simulation is used in the airline industry to teach new skills to pilots or refresh existing skills. Simulating a crash situation is obviously less expensive and much safer than actual experience.

*Systems modeling.* Systems modeling recreates or models a real life system so that students can experiment by changing or modifying parameters surrounding the system and observing the varying results of the changes made. One example is experimentation on ecological systems. Introducing or removing hazardous chemicals into or from a simulated environment can save many hours of courtroom time as well as heavy fines.

*Experiential simulation.* The third type of simulation, experiential, provides a series of experiences or procedures that would not normally be available to the student. The steps involved in preparing for foreign travel, for example, can be presented to the student before such traveling occurs, again without the associated dangers of actually being thrust into the situation.

All three types of simulation—performance, systems modeling, and experiential—are very closely related and may overlap.

### *Problem Solving*

Problem solving is closely related to simulation. The difference is that the student uses the computer to find solutions to problems that would take too long to discover manually.

Problem solving takes place on two different levels. The first level uses existing programs and known algorithms or equations to find the solution to a problem. The student enters the variables into a fixed equation and then lets the computer do the work and produce the answer. This allows the student to concentrate fully on the problem-solving process without getting sidetracked with unnecessary involvement in the mechanics of the process. Trainers sometimes use this method when performing statistical analysis on student groups. Calculation of standard deviation is an example.

The second form of problem solving allows the student to generate his own algorithm. That is, manipulative factors as well as the variables can be defined. Spreadsheet programs accomplish this by allowing the user to choose and enter not only the data to be manipulated but also the manner of manipulation.

*Informational*

This type of CAI program is designed to deliver information such as rules, data lists, instructions, or procedures. The major characteristic of this format is that it involves very little interaction with the student. The student must initiate an inquiry with the computer, but then the computer simply delivers the information requested with no further student interaction. Informational programs are used extensively in computer-managed instructional environments (CMI) to deliver instructions to the student and guide the subsequent steps of the training program.

*Instructional Gaming*

Instructional gaming differs from the typical video game concept in that instructional games call upon the student to apply specific skills. The games have very specific objectives intended to help the student learn critical skills needed on the job. They are important to mention because of their built-in self-motivational aspects. In other words, while their objective is to satisfy the specific needs of business, they are fun too. Instructional gaming is closely related to problem solving and simulation methods, but is much more interesting.

## COMPUTER-MANAGED INSTRUCTION (CMI)

Computer-managed instruction, or CMI, uses the computer to make efficient use of limited instructional resources. These resources are defined as materials, personnel, equipment, and time. Unlike CAI, the student is not totally involved with use of the computer and there is often outside intervention involved in the learning process.

CMI can be divided into two subapplications. They are

1. Instructional Management
2. Administration and Support

*Instructional Management*

Instructional management is the second most visible form of computer-based training (after CAI). Instructional management utilizes the computer in the classroom to direct the learner through a series of exercises in order to learn the subject matter. Exercises may include other forms of media, live instruction, or CAI programs. The methodology and exercises must be determined and planned in advance. The instructional plan derived is then programmed into the computer. As the student progresses through the instructional plan, the computer can also track student progress, test, deliver instructions at various points, and direct the student to other resources.

*Administration and Support*

Administration and support CMI is often overlooked or goes unrecognized as CMI because the student's direct involvement with the computer is minimized. From the instructor's point of view, however, it is perhaps the most useful and powerful application of the computer in the instructional setting.

Programs for the administration and support of training include

1. test generation
2. record keeping
3. scheduling
4. statistical analysis
5. report writing
6. resource cataloging/location
7. instructional product evaluation
8. media and materials design

## TYPES OF SOFTWARE AVAILABLE

There are four primary ways to attempt instructional delivery or management using the computer. They are

1. Write your own programs.
2. Use another program to create a program.
3. Purchase prewritten programs.
4. Contract someone else to create a program for you.

*Writing Your Own Programs*

If you want to write your own programs, you will find it very time-consuming and you had better be well versed in the programming language that you use to write the program. There are literally hundreds of programming languages on the market to choose from. The most commonly known is BASIC (Beginner's All-purpose Symbolic Instruction Code), which was discussed earlier. Its use invokes the execution of machine language instructions to accomplish the task defined in the higher level language statements of the BASIC program.

Programming languages like BASIC and others are satisfactory for general program writing. Each of these higher level languages has its specific strengths and weaknesses. Very few of them, however, have the specific statement repertoire to generate efficiently written CAI or CMI programs. That is, many of these higher level languages, while excellent in their own right, are not well suited to the specific concerns and issues surrounding the creation of CAI/ CMI programs.

One of the more familiar higher level languages available for use in CAI/CMI applications is PILOT (Programmed Inquiry, Learning, Or Teaching). It is a programming language designed specifically to aid in the creation of CAI and CMI instructional support software. PILOT is designed to help you write programs that will handle such situations as incorrect responses, misspelled words, synonyms, branching, and evaluation. Keep in mind that other higher level languages can be used in the instructional environment, but perhaps not as efficiently.

In short, while languages such as PILOT are available to aid you in the creation of instructionally sound and efficient software, the concept of writing your own program may not be the most cost-effective way to produce CAI/CMI training materials. Certainly in industry, time is money. This method should be reserved for the creation of programs that will, by their nature, produce a satisfactory return on investment.

### *Using Other Programs to Create a Program*

A more efficient and time expedient way of dealing with the problems of creating instructional materials on the computer is to use existing programs to create a training program that is useful to you. One of the most popular ways to do this is to use an *authoring* program. An authoring progam provides a specific framework in which to enter data or information for delivery to your students. There are many authoring programs available to suit a wide variety of needs and budgets.

Simple authoring programs are inexpensive but, as you might suspect, are limited in their capabilities and applications. These programs are usually linear in nature and therefore have limited branching capabilities. They usually deliver information and test for comprehension only. The more sophisticated authoring systems also have the ability to generate specific feedback to the student and some data for the instructor, so that both can track the student's progress.

There are still more sophisticated authoring systems available that allow you to produce CAI programs with extensive branching options, testing mechanisms, and remedial capabilities. Some of these systems allow you to interface with video cassette recorders or tape/slide machines, so that you can incorporate variety of delivery into the training program. Needless to say, authoring programs in this category are more expensive. They are, though, one of the easiest and most cost-effective means of producing CAI/CMI programs that meet the needs of the training institution and the student.

### *Programs Often Overlooked*

Authoring software is good for developing CAI and some CMI programs; however, the programs often do not fully support the administration and support function in the training environment. There are a number of programs available that do support training program administration and support, and they are often overlooked by many trainers and training administrators.

*Spreadsheet* software, for example, can be used for record keeping, statistical analysis, scheduling, and evaluation as well. These kind of programs will manipulate figures easily and allow trainers some facility in evaluation in particular. An example of the use of spreadsheet programs in the administration and support function is shown in figure 8.4.

Some of the more sophisticated spreadsheet programs can be used to analyze test scores, calculate the mean, median, and mode scores in a class, and calculate the group standard deviation. Some also have the ability to graph this information so that subsequent analysis is made easier.

Another often overlooked program is the *data base generator*, or *data base manager*. The data base generator, or manager, creates a data base, or "pool" of information and data, which can then be modified or added to and formatted into report format. The user is responsible simply for entering the data and defining the report format. These programs can be used for test generation and materials selection. They are also used to catalog and later locate instructional materials.

If you are typical of most trainers, you probably have numerous references that are used in a variety of training applications and programs. With a good data base program, you can file information regarding these numerous resources and later retrieve listings of resources keyed to a particular common criteria. By doing this, much time can be saved in the location and selection of appropriate resources.

The generation of bibliographies is a perfect example where use of a data base program can save time and money. How many times have you wished that you could find the time to generate a bibliography for a training program that will probably only be given once or twice? By using the computer to store bibliographical information, a list is always ready. All you do is indicate the search criteria to the data base program. The program then

EXAMPLE

| Name | Grade 1 | Grade 2 | Grade 3 | Average |
|---|---|---|---|---|
| Student 1 | 79 | 68 | 88 | 78.3 |
| Student 2 | 65 | 79 | 95 | 79.6 |
| Student 3 | 88 | 90 | 98 | 92.0 |
| Average | 77.3 | 79.0 | 93.6 | |

*Figure 8.4 Spread sheet.*

searches the entire volume of information stored and prints out a list of all data items that satisfy the criteria indicated.

Do not forget *word processing* programs. Most of them are extremely flexible, easy to learn, and can generate training documents tailored to individual sessions. Once documents have been created and entered, they can be stored on a disc medium for later recall. Then the needs of similar future training programs can be quickly met by loading the original document, making appropriate changes, and then printing the "new" document. The new document can also be saved on a disc for later use.

One form of word processing program that is often overlooked is one designed for processing large documents, such as training manuals. This type of program can incorporate graphics and written material on one page, using the same program. Such programs can save hours of cutting and pasting. If the material that you present changes frequently, the use of word processing is a must.

All authoring or other support programs discussed so far will take some time to set up. The majority, if not all, of the programming tasks have already been done for you, but you must still establish and enter the operating parameters for a particular system, define equations, and enter the appropriate information and data.

### *Use of Purchased Programs*

Purchased programs differ from prewritten authoring programs in that they are "canned" and ready to use in a specific instructional environment. Initial preparatory work on your part is held to a minimum. These programs would include tutorials on popular existing programs, general information programs, and ready-to-use record-keeping programs.

Tutorials can be purchased for most of the spreadsheet programs available as well as some of the data base and word processing programs. Training programs already exist to teach such things as computer literacy, typing, programming, and other computer-related skills. In time, rest assured that other programs will be developed for delivery on the computer. The National Education Council is already developing training programs at the college level that can be taken at home. They are administered using your home computer as the course facilitator. Written materials and televised programs on public broadcasting stations will support the courses, which will be sponsored by local colleges and universities.

Other prewritten programs already exist that will support the keeping of records and generation of tests. With these menu-driven programs, you need only enter the data. All of the equations and many preestablished formats are already built into them. Tests can be generated from a data base of questions and scored by entering the answers for each student into the computer. Some of the more sophisticated programs will even analyze a group's test scores automatically. This analysis is useful later when trying to identify and correct weaknesses in delivery of certain portions of the course.

*Media generators.* Media Generators are a little-known form of prewritten software that are nonetheless available today. These programs allow you to use the computer as a graphics design station to create visuals for use in training programs. Most programs of this type are available from vendors who can support their conversion to 35-mm slides or overhead transparencies. All you need to do is design the visual, save it on a disc medium, and deliver the disc to the vendor. The vendor will then enhance the image resolution and produce the slides required.

This type of prewritten program can save hundreds of hours a year, especially if you use large volumes of slides or transparencies in training.

### *Contracting Someone Else*

The fourth option available is to contract a vendor to develop the software for your particular training requirements. The vendor's work will result in a program that you simply load into the computer and run. Although more expensive initially, this method is cost effective if the program under development will not change appreciably for some time to come and if it will be distributed to a large target audience that cannot be controlled by the instructor.

At present, custom software houses are springing up to accommodate a vast number of applications, including the support of training. Software vendors who support training can do everything for you, from writing a simple drill and practice program to producing an interactive program that interfaces with other instructional media such as video.

## HARDWARE

One aspect of CBT that we have yet to address is the hardware required to implement any of the systems already discussed. The computer that you use should not be a big factor in the implementation of CBT. Any of the programs mentioned thus far can be found to run on many of the popular personal computers available. If you are serious about setting up a CBT system, you should be primarily concerned with your long-term needs and pick the system of hardware and software that best serves those needs.

Some of the questions you will need to answer prior to choosing the hardware components for your system include

1. How much memory will I need? (Check memory requirements on the software that interests you.)
2. What type and how much mass storage (disc, etc.) will I need?
3. Will a number of terminals be used, and will time-sharing be needed?
4. Do I need color?
5. Will I need hard copy (printout) capability?
6. Will I need sound? (Not all computers have sound capability.)
7. Will I eventually want to interface with video or slide/tape?
8. Will I eventually want to use specialized peripheral equipment such as joysticks, light pens, or touch screen? (See Chapter 9.)
9. How much money am I limited to, if applicable (and it probably is)?

Some of the less-expensive computers available are appealing at first, but as you grow and try more sophisticated techniques, you may find that each of those techniques that you want to implement will require an additional piece of hardware. In the end, the less-expensive models cost more than had you opted for a midrange-priced system to start with, which would have many of the desired features built in.

No attempt will be made in this text to single out any one system. The choice is entirely up to you, and probably more than one system is capable of performing to your specifications anyway. As you enter the realm of computers in training, you should talk to a consultant who is well versed in the application of computers in the training environment. Such consultants will be able to help you analyze your particular situation and help you make a decision about which system will work best for you.

Any recommendations you obtain should come from unbiased sources. This excludes vendors. Recommendations or advice should come from trusted colleagues, although you should remember that the recommendations of colleagues are based on their experiences and the use of the computer in their environment, which may not be the same as your environment.

*Looking for Hardware*

While variability is always a consideration to remember, the basic hardware needs for one CBT station usually include

1. the computer itself (CPU)
2. a disc unit for storage
3. a CRT display
4. a keyboard for input
5. a printer for hardcopy

This hardware is considered standard for CBT workstations and necessary for realizing a flexible and powerful system.

If you already have this equipment, you may be ready to expand upon its capabilities. If you are just getting started, you may wish to consider how you will utilize the system. Either way, be aware of what peripheral devices to look for and how they can be used in training, and make sure that you need them before you buy them. Keep in mind that hardware is inherently "dumb." Successful implementation of hardware, therefore, is a function of being able to create or find (purchase) the software that will make such hardware "smart." Chapter 9 will discuss some of the peripheral devices that are commonly found on computer-based training systems.

## SUMMARY

There are two major categories of computer-based training (CBT). They are computer-aided instruction (CAI) and computer-managed instruction (CMI). Computer-aided instruction embodies those concepts pertinent to the delivery of instruction. Computer-managed instruction involves the administration and support of courses, whether they are delivered on the computer or not.

Within the realm of CAI, six applications can be identified. They are drill and practice, tutorials, simulation, problem solving, informational delivery, and instructional gaming. Three types of simulation are performance simulation, systems modeling, and experiential simulation.

Computer-managed instruction encompasses instructional management, which serves to direct the learner, and administration and support mechanisms, which test, keep records, schedule, analyze, write reports, catalog, evaluate, and design media and materials.

Software, of course, forms the heart of a computer-based training system. Software is available through several means. These include writing your own software, which can be very time consuming; use of authoring programs, which are specifically written for instructional design; use of purchased programs, which are ready to implement in specific situations; and contracting someone else to write software for you.

When scrutinizing hardware, consideration should be given to the amount of memory required, the number and types of peripheral devices needed, and other factors involving color, sound, and the need to interface with other types of delivery equipment, such as video or slide/tape. Money is almost always a factor as well. Successful implementation of software is vital to the success of the system, since hardware without software is inherently "dumb."

# Objectives

By the end of this chapter you should be able to

1. state the factors that will affect your choice of CBT strategies
2. state the benefits of using the computer as the delivery system for training
3. identify the peripheral hardware needed to implement specialized CBT applications

# 9 Beginning to Apply Computer-Based Training

## INTRODUCTION

Chapter 8 discussed the basic terminology and uses of computer-based training. It also examined the basic hardware and software components that comprise a system supportive of computer-based training situations.

The next consideration is that of when to use CBT as your delivery or management tool. Keep in mind from the start that the computer is not always a practical or cost-effective method to employ in training and development. If used incorrectly, it can actually detract from the effectiveness of training in given situations and/or add to the cost of delivery. Also, the training cost per student may actually increase over that of conventional methods. Therefore, its use is not always cost effective.

This chapter will determine some of the criteria that you should examine and test prior to making that sizable initial investment in equipment. The data that you gather as a result will generally indicate whether or not the computer is the appropriate tool for use in your particular training application.

## CRITERIA FOR USE OF CBT

Computer-based training can be an extremely cost-effective method for the delivery of high-quality and uniform training if certain criteria are tested and met prior to its implementation. Consider the following factors before you decide to use it.

*Cost Effectiveness*

Very often in business and industry, as well as other sectors, cost of implementation is a key factor in deciding whether or not to purchase any item for the business. This certainly applies to the purchase and use of a computer system as a delivery and/or instructional management tool.

Many times, company training programs are requested, designed, and delivered to the requesting department, and never or seldom ever given again. This is due in some cases to the specialized nature of the program, which makes it unsuitable for distribution to most of the potential users of training in a given environment. In this situation, the cost to develop this "one-shot" program on the computer would obviously be high and as such would not favor using the computer at all.

Specialized training courses, for which no program or programs already exist, will take a considerable amount of time to develop on the computer if effectiveness of the final product is to be maintained. If training departments that operate as profit centers did develop programs falling into this category, the resulting costs transferred out to the requesting department would be abnormally high. The requesting department, having been duly consulted with regard to these costs ahead of time, may well choose to defer such costs. The result, of course, is that training "business" is lost or at best deferred.

The training department in any company exists to serve the educational needs of other divisions and departments of the company. Promising an excellent program to the manager of another department carries little weight if, in fact, the associated cost to that manager's budget is prohibitive. Managers are judged in part on their ability to spend budget allotments carefully and to stay within budget. Often, the exact costs to implement programs are not known to the requesting manager at the time budgets are prepared, although it may be known that a program will be needed the following budget year. The training department, which is aware of the manager's need, may also not know what the associated costs will be since development of the program may not have begun.

Few managers will give an authorization to proceed with development of the training program(s) needed until they have an accurate assessment of the costs involved. This problem is common and the best solution is for the training department to maintain a constant vigil on its costs to develop programs recognizing that the "buyer" will not buy if the "product" is overpriced.

Training departments that do not try to recover their costs through budget transfers face the same scrutiny by their own upper management, which is always conscious of the operations costs of the training department. After all, this cost represents an overhead expense of doing business and as such impacts profitability. In cases where training is to be given once or twice to few students in a specialized environment, noncomputer development may be the most prudent approach.

### *Timeliness*

Many managers often want a training program "yesterday." The time frame that they operate under may dictate the need for training as soon as possible. This obviously places a burden on the training department to design, develop, and deliver the needed instruction quickly. Again, if no library programs (programs that already exist) are available, the use of the computer should not be considered because the time required to develop that program may far exceed the requesting department's specifications. If it is great, but late, the overall effectiveness is lost. "Cut and paste" technology may be the best answer here.

### *Subject Matter*

Computer-based training can be used to deliver a wide variety of topics. Future modifications to the subject matter to fit changing needs can also be done quickly and effectively, as long as the changes are relatively minor. If the subject matter is very volatile and therefore subject to frequent or extensive modification to keep it current, use of the computer system as a delivery agent will not be cost effective.

Even after a program has been initially designed, developed, and safely stored away on a disc device, the cost to deliver training thereafter is affected by an additional factor known to all of us as course maintenance. Changes made to existing computer-based training cost as much per hour, if not more, than the initial costs of development and implementation. Frequent or substantive changes to courses, brought about by changes in circuit design, for example, keep the cost of delivering that program high and therefore undesirable in the eyes of the budget-conscious user.

Use of the computer as a course management tool (CMI) may also suffer from the effects of frequent or substantive course changes, even if the delivery tools were developed manually. Suppose, for example, that students worked through modules that were manually developed but then took a criterion test (or tests) administered and graded on the computer. Suppose further that such tests were subject to frequent modification due to changes in the subject matter, which in turn influenced the criteria for successful completion of the course. If you use the computer to perform statistical analysis on the grades of the entire class, associated analytical software may also require changes. Thus, a "domino" effect might occur that could translate to a significant amount of time required to modify the course administration mechanisms supported by the computer.

### *Target Population Size*

Another factor that should affect your decision to use computers is the size of the target population that you intend the program to reach. Materials cannot be developed effectively on the computer for very select groups or limited numbers of people and still maintain some reasonable limit on the cost to develop the program.

An example of the effect of target population size on the development cost of a course was cited in a previous section. If a very specialized program for a total of six students were required, and the development time for that program was on a ratio of 150:1 (150 development hours for one hour of instruction, which is not uncommon), the total development time for such a program could total several years, depending on the length of the training program. Obviously, this kind of cost involvement could well prohibit course development to begin with. The labor costs alone would be staggering, not to mention the cost of the hardware, software, and so forth.

To represent a cost-effective training tool, then, and given that the development ratio above is probably conservative in many situations, the computer should be used as a development and delivery tool only in situations where the resulting course material will be delivered to a large number of students over a reasonable period of time. Only in these circumstances will the computer lend itself to cost-effective development and delivery.

You may be thinking at this point that you should forget using the computer at all just because you work in a small company that does not train large numbers of students in any topic. The size of your business alone may not be a factor. If your company experiences a high turnover for example, and the subject matter you are considering tends to remain the same and is applicable to all new employees, the "quantity of students" consideration implied in this section may be met in your situation as well. The cost of "canned" new employee seminars or basic training administered to all employees may still be minimized through the use of the computer, particularly if a high percentage of personnel involvement would otherwise be necessary to carry out this training requirement. A growing organization could well benefit from the use of the computer to train new employees hired.

### *Student Location and Distribution*

Delivery of training to the student as opposed to the student coming to a central training facility is an ideal situation for any company or instructor. Use of the computer facilitates this process because it can be used to train students in remote locations away from the central processing facility. Several companies come to mind which now employ rather sophisticated instructional delivery and management systems for use in training service technicians in a number of different branch locations. Implementation of this approach eliminates, in many cases, the time and expense involved to transport a student to a central training facility. In larger firms particularly, this could result in a substantial cost savings over time.

This method of "remote delivery" is not always cost effective as you might already suspect. Factors involved here include the number of remote delivery points as well as the number of students to be trained at each point. For example, the expense involved to equip a branch office that only plans on training three students per year using that equipment may not be justifiable. On the other hand, if that branch had plans to train 100 students in a given program during the first year of operation, this could well be justification enough to spend the money on remote equipment in the branch.

Another factor in the decision involves the length of the training program to be administered in the branch. The cost to send a student to a central facility to attend a one-day training program might be substantially reduced if that program were offered effectively in the branch on a remote terminal. On the other hand, the overall cost to send a student to a four-week training program may be only marginally reduced by offering that course in the branch office. In this case, the cost may not be reduced at all.

Still another factor to consider is the number of delivery points to be implemented. There is obviously an incremental cost involved in equipping and operating added branch training facilities. The decisions made in this dimension should be tempered with the number of students from each such branch that would be considered candidates for training on the equipment employed. Ideally, cost effectiveness is most likely achieved in the situation where a number of candidates for training reside in each of relatively few branch offices. This situation would maximize the use of minimum capital equipment dollars invested for needed equipment.

The warning discussed above, regarding the delivery of highly specialized training designed for a select few candidates, should enter into the decision here as well. Truly then, "price/performance" applies. Price in this scenario is defined as the cost to implement a remote (branch) training facility. Performance is defined as the amount of on-line time (actual training time) realized in each branch as a result of implementation.

One final consideration in this section deals with the type of training to be delivered. If the training programs being considered for delivery by remote means would normally involve a high degree of direct interaction with other training aids, then the cost to equip each branch with such training aids must also be considered. Obviously, in any training situation or location, the student must have access to all of the necessary support equipment required to effectively dispatch the training at hand. In technically related disciplines, for example, students may be required to adjust, take measurements on, or troubleshoot highly sophisticated equipment in order to satisfy the course requirements. The cost to provide such training effectively in a remote or several remote locations may skyrocket due to the added cost of such equipment.

Possibly in such situations satisfactory completion of programs by the student can only be assured through the use of a proctor (instructor) as well. Now you're looking at the cost of the computer terminal equipment to deliver the training, the cost of capital equipment on which the student can work, and the cost of a proctor to monitor, for example, the student's electromechanical or electronic acuity. All of a sudden then the picture changes. It may be advisable to transport the student to a central training facility and forget the concept of remote training by computer. Indeed in such situations the company may well save money by establishing and equipping a central training facility. (Note: the concept of price/performance still applies.)

### *High Interaction*

One of the primary objectives of effective computer-based training is to involve the student in the training process as much as possible. Training done using a computer as the chief delivery medium is most effective if the student is involved. Simply stated, if long periods of time will tend to lapse between student responses, or if the student will be subjected to page after page of material with no required involvement, then computer-based training will not be the most effective method for delivering that training.

A computer is many things and can solve a wide range of training problems, but it is not really an inherently personable instructor. Student interaction and involvement must be maintained at a high level in order to maintain the student's interest in the material being taught. You would not use video tape as the sole delivery mechanism for a two-week, eight-hour-per-day training program. Computer-based training relies heavily on the video medium as well. Do not use it at all unless you can conveniently plan on and implement regular student interaction. Course delivery should be done in small modules with regular checks for comprehension requiring student response.

## COST AND QUALITY CHECKLIST

After reviewing the criteria in section 9.2, you might well decide that computer-based training may be an appropriate method in your situation. If so, then the following benefits should prove to be true:

1. The amount of time that it takes to train a student will be reduced.
2. Courses will be made more instructor independent, freeing the instructor to perform other tasks.
3. Overall costs to deliver effective training will be reduced over alternative methods.
4. The cost of ongoing course maintenance will be reduced as compared to alternative methods.
5. Training delivered by computer will be of consistently high quality.
6. Training, by nature of the program, is applicable and therefore distributable on a large scale.
7. Effective and cost-effective training can be distributed to remote sites. (This may not be applicable.)
8. Instruction will be highly interactive and performance based.
9. Instruction can be modified or individualized with a minimum of redesign.

## PERIPHERAL DEVICES IN CBT

The following text defines and describes the use of some of the more unusual peripheral devices that have found particular use in training and development.

### *Card Readers*

Card readers use optical scanning equipment similar to the ones that are showing up at supermarket checkout counters. They come in a variety of sizes and can read standard IBM cards or full 8½-by-11-inch pages. Using card readers with test generation and scoring programs makes objective evaluation a snap. Imagine having the tests for a group of 20 students scored, analyzed, and plotted during a 10-minute break. Feedback can be almost instantaneous. Individual students can utilize the card reader to grade their own tests while taking a self-paced program.

### *Light Pens*

Light pens are used most often for responding to information presented on a CRT. Choices are displayed on the screen and the student simply puts the light pen on the screen to indicate the answer. The computer software responds positively to areas of the screen where the correct answer is displayed and prompts the student if an incorrect answer is selected. This device allows you to present training programs using CBT to people who have little or no background knowledge of the computer or keyboarding skills.

### *Bar Code Readers*

One variation of the light pen is the bar code reader. This device and its use have an advantage over the use of the light pen and CRT in that bars can be placed anywhere on the screen without regard to "inserting" their location in the software. The student simply runs the bar code reader over the series of bars on the screen. Bars can also be scanned from a sheet of paper, thus facilitating question and answer exercises on hard copy as well.

### *Plasma Screens*

The plasma screen, commonly referred to as a "touch screen," allows you to do away with a light pen altogether. A sensing device is built into the CRT and connected to the computer using a system of coordinates. The computer software responds similarly to the location touched by the student's finger, as to the light pen. Plasma screens are available for many of the popular personal computers.

### *Digitizing Pads*

Digitizing pads are more an instructor's tool than a device for the student. With a digitizing pad, the instructor can transfer graphic information from hardcopy to CRT screen. This allows fast generation of computer graphics and is relatively easy to learn.

*Key Pads (Responders)*

Key pads, sometimes known as responders, can be used as individual units or tied into a network for use by a group. As used individually, the instructor can minimize student's initial fear of the computer by eliminating the full keyboard. The student is given a device that looks like a hand-held calculator or television remote control. All student responses are numeric or are made by pushing "function" buttons.

Responders can be used in group sessions also. Each student would be equipped with a responder. By linking all of them into the computer, you can deliver a live presentation, then ask students questions for comprehension as you go, with the students responding with their hand-held units. Answers are immediately tabulated and summarized so that the instructor can spot any weak areas in training in real time, and schedule review sessions accordingly.

*CRT Cameras*

Many times the computer is not the best delivery system to use in a given training scenario. At the same time, it has the capability to easily produce colorful graphics. A problem arises when you want to show such graphics to others. One solution is to invest in a CRT camera. The CRT camera takes the signal directly from the computer, enhances it, and allows you to photograph a slide of it. If you are producing great quantities of slides each year in support of training programs in your setting, this could be one of the best time-saving devices you could buy.

At present, there are models available to produce everything from 35-mm slides to 8-by-10-inch overhead transparencies. You should know however that the price of a CRT camera can range from $2500 to $80,000. The key to investing in a CRT camera lies in the quantity of slides and such that you produce in a given time period and therefore the money that would be saved by such an investment. Only then will you know whether or not the investment is feasible.

*Interfacing with Other Media*

Interfacing other media with the computer is becoming more and more popular. Video, using a video disk player, is an often-talked-about option. It is fast and fairly easy to use once the original video is made and transferred to the video disk. This option, however, is also fairly expensive. If your subject matter is not subject to change, it's perhaps the best method. If you cannot afford video disk systems, look into using video cassette decks. They are somewhat slower in implementation but also less expensive. With the right software, you can produce programs that use them yourself.

Do not overlook using slide/tape programs with your computer. Many slide/tape players available now are made to interface with a computer so that you can present a short segment that ties in with a CAI program.

## SUMMARY

This chapter has discussed some of the considerations to be examined prior to making the decision and associated financial commitment to use computer-based training. As you have no doubt realized, the responsible decision to implement computer-based training requires examination and planning and should be initially approached on an objective basis (after analysis, you may determine that computer-based training is not the wave of the future in your situation).

While we observe this statement somewhat simplistically from hindsight now, you should know that in the past companies have poorly implemented computer-based instructional methods. They have spent more money and/or achieved less because their approach was not well thought out. While the guidelines cited in this chapter may at first seem overwhelming, their examination really involves a lot of common sense as well as expertise in a technology. Only after you have made the decision to implement computer-based training can you use the information in Part I of this text to continue the process toward smart selection and implementation of the system.

# Objectives

At the end of this chapter you should be able to outline three standard applications of the computer in training.

# 10 Case Studies in Computer-Based Training

## INTRODUCTION

Where do we go from here? By now you should be familiar with the basic terminology and uses of computer-based training and have a rough idea as to when or when not to use it.

This chapter will outline three hypothetical applications of the computer in training:

- where equipment is available or in place
- where courses are in effect on an ongoing basis
- where there is an extremely limited budget

These three cases are general in nature in order to use them as models in a variety of other situations.

## CASE 1

Case 1 involves a company that already has computers available. The company may make computers as a part of their product line or have a network established for conducting company business. Their computers are general purpose microcomputers that can be used for virtually any application. At present the company uses them for word processing, financial transactions, and data base management.

Training is a problem due to the high turnover in personnel in critical areas such as sales. The product line is extensive. Many of the products were developed over 20 years ago but are still revenue generators today.

Training must be delivered in a timely manner. Personnel who must wait three to six months for product training will not be very productive in the interim. The chances of these people receiving all of the information that they need on the job are low. The extensiveness of the product line makes it impossible to teach every person about every product. In addition, the target population is scattered worldwide. Finally, newly hired personnel do not arrive on board at the same time. The cost effectiveness of training individuals on a one-to-one basis is poor.

The conditions described here would indicate that some level of CBT could be implemented successfully. Hardware is not a problem since every office has at least one microcomputer already installed. Management has agreed to allow time on the computer for training purposes.

The major problem in this case is the organization and development of CAI modules. Implementation of a training system could begin by developing stand-alone CAI modules for each product. If training needs to be developed quickly, menu-driven authoring programs can be used. In time, more specialized CAI progams can be written. A plan of instruction can be generated for each student based on the number and types of products that need to be learned and when the knowledge is needed. This plan can be generated by a planning program housed in the computer. Students can be evaluated by administering criterion tests at the end of each learning module.

The CAI modules can also be used as job aids after the initial training has been completed. Employees can use the training materials for review virtually anytime by using one of the terminals on the network. Salespeople can review a product's key features prior to making a sales call or when planning for the product's use in a specific application. If the CAI modules are carefully written, so as not to compromise company-confidential information, they could even be sold to other firms in the industry over a national computer network. This could defray the cost of producing additional needed modules.

Due to the criterion-referenced nature of training suggested, the time to train a new employee could be greatly reduced. The employee becomes more productive sooner than would otherwise be possible. Initial cost for appropriate software and development of initial modules must be considered in this case, as in any other case. The prudent firm would obviously benefit from the performance of a rigorous cost/benefit analysis prior to expending the funds required.

## CASE 2

Case 2 observes a situation where training or education is an ongoing process. Such situations exist in universities, colleges, and in technical training organizations within industry. The "product" of these organizations is training. Formal courses are presented to a group of students. There are always enough students available to hold a class. The classes center around skills training and are therefore heavily performance based.

In this situation a problem exists in the student body. The students have wide-ranging prerequisite skills. They come from a variety of backgrounds which are not controllable by the instructor. Students, nonetheless, are paying for the course and expect to learn as much as they can in that course. It is the course instructor's responsibility to teach them what they need to know in order for them to be able to perform their job capably when they leave the classroom. If the course is aimed at the students who are weak in the prerequisite skills, those who are more proficient will complain that the course lacks the proper pace. If the course is aimed at the more highly capable students, those with weak entry skills will suffer and learning may not take place. CBT can help, by allowing the instructor to tailor the course design and pace to each student's specific capabilities and background.

Another problem exists, however, in that departmental budgets will not allow the purchase of terminals for every student. One answer is to invest in enough terminals to accommodate about 20 percent of the students at any one time. In a course with 30 students, this would amount to six terminals. Since the program of instruction is skills and performance based, the students will be performing practical exercises most of the time. Not all of the students will be working on the terminal at the same time, particularly if the course can be designed to be criterion referenced. Criterion-referenced design would also eliminate the problems inherently found when lecturing to a student population with diverse prerequisite skills. The terminals would be more heavily used by those students needing more prerequisite information.

All students entering the course can be tested to determine their entry level or background skills. Those who are weak can be directed to CAI programs. A "prescription" tailored to the individual can then be written for every individual taking the course. The students can progress at their own comfortable rate from one practical exercise to another. Testing can be done on the computer and records kept for evaluation. In short, terminals can be shared for remediation, testing, and delivery of supplementary information. Lecture sessions can still be held if they are scheduled. Students can then budget their time accordingly to allow attendance.

Implementing a plan such as that described above can also free the instructor to spend additional time with those students requiring it, or discussing more advanced concepts with those students who are ahead of the group. The latter situation is often a luxury that the instructor seldom enjoys.

## CASE 3

Case 3 is probably the most common situation existing in training departments or institutions of higher learning today. Before we examine it, take a look at your own situation first.

If you are tasked with the job of training, you must develop a plan for the delivery of that training and the development of the materials to support it. The students that you teach must also be evaluated in some way. Results of these evaluations may need to be organized in a formal report. You may be tasked with keeping track of the students' training history, particularly if they are employees of your company. Course schedules must be kept and maintained as well as lists of students to attend each course.

Facilities and resources are also scheduled and budgets must be monitored.

The organization in Case 3 is tasked with ongoing training of personnel on an extremely limited budget. The instructors are also developers and must generate and maintain all of their training materials, including tests, manuals, workbooks, and practical exercises. Training administration involves the scheduling of rooms, students, and resources. Budget monitoring is close and constant due to limited availability of funds.

With all of this to do, minimally the operation aspects of training as opposed to administration, the instructor's time is taxed to the limit, or so it seems anyway. Many times, student evaluation is skimmed over quickly in order to ensure that all information is delivered within the time allotted. Course materials and tests are not updated properly because of the time involved in retyping pages or whole documents.

Case 3 is another prime candidate for using the most overlooked aspect of CBT, the use of the computer for training support. Cases 1 and 2 dealt with the use of the computer to help the student in the classroom. Case 3 helps the student, but does it indirectly by saving the instructor's valuable time and improving the quality of training materials. Instructors who use a computer to manage their courses and generate their materials can save up to 50 percent of the time usually spent on the following tasks:

REGISTRATION: Information can be entered directly into a computerized data base to register students for courses. This information can also be used to generate reports or analyze the student population at a later date.

RESOURCE SCHEDULING: Resources such as rooms and equipment can be scheduled using the computer. The chances of double-booking or not having enough equipment will be reduced. Reports on utilization of resources can be generated easily at any time.

MATERIALS DEVELOPMENT: Course descriptions, plans of instruction, lesson plans, training modules, exercise sheets, and other documents requiring design and update, can be done on a computer much faster than by using an eraser, pencil, and typewriter.

TEST GENERATION, SCORING, AND EVALUATION: Test questions can be written and stored for future use using a computer and a data base type program. Generating a new test becomes a matter of stepping through the program and identifying the questions that you want to use. Scoring can be as simple as entering each student's name and answers. Evaluation of scores or of each question becomes almost automatic, with all data manipulation and arithmetic operations being done by the computer.

INFORMATION FILING: Slide files, collections of overhead transparencies, video tape libraries, and bibliographies of printed materials can be stored in a data base. Finding the information that you want to use becomes easy and quick.

STUDENT RECORDS: If you are training employees of your own organization, you may be tasked with keeping training records on them. The computer can keep historical records on students and help generate a report when necessary.

The increased efficiency in performing all of the above tasks and the savings of time realized over the first year or two alone should help to justify the purchase of the computer and needed software to support the training function.

# Glossary

ACCESS TIME: The time interval between the instant of a request for data from memory by the CPU and the instant that data is retrieved. Also, the time between the instant of a request for data from a disc and the beginning of retrieval of the requested data.

ADDRESS: Designation of the location of data within the memory of a computer system.

ALU: Arithmetic and logic unit—the portion of a central processing unit that performs arithmetic operations.

ASCII: American Standard Code for Information Interchange—a convention for coding the characters that can be generated by a typical computer and printed by a printer or CRT. The standard set includes codes for letters, numbers, and special characters.

BACKUP: A duplicate disc of important data and programs stored as insurance against the possible loss or destruction of an original disc's data.

BASIC: Beginner's All-purpose Symbolic Instruction Code—a higher level language common to most microcomputer systems.

BINARY: A number system using a positional framework and only two numbers, 0 and 1, to represent large numbers. It is compatible with the general architecture of digital computers as they use only two voltages (high and low) to represent the presence or absence of facts, states, or conditions.

BIT: *BI*nary digi*t*—a portion of a byte; the smallest unit of data storage with a value of either 0 or 1.

BUG: A problem existing in a program which prevents the program from achieving its desired objectives. Programs with bugs need to be debugged by the programmer.

BOOT: The process of loading a computer's memory with software initially so that it can go on to perform other tasks; as in booting a BASIC operating system so that programming in BASIC might take place.

BUFFER: A storage area in a printer, for example, which allows faster transfer of data from a faster device, such as a computer, to a slower device (the printer), thus freeing the computer to work on other tasks. The slower device is now in a position to work with (print, display) data independently of its "master." Any output terminal may be buffered.

BYTE: A unit of measure in storage; for example, a memory may have 32K bytes of storage capability. Each byte in a given memory is comprised of a certain number of bits.

CAI: Computer-assisted instruction—the direct use of the computer to assist, enhance, or replace the instructional process. CAI can include software programs described as drill and practice, tutorial, problem solving, simulations, gaming, or informational.

CBT: Computer-based-instruction—term which globally encompasses the use of the computer in training.

CHARACTER: A number, letter, or special symbol on the computer's keyboard, or found in printed or displayed text.

CHIP: The result of microminiaturization of logic circuits, printed in layers on a very small rectangle of silicon material.

CMI: Computer-managed instruction—the use of the computer to support the instructional process. CMI can also include software programs designed for producing tests, record keeping, monitoring of learning objectives, diagnostic or prescriptive functions, and resource management.

COBOL: COmmon Business-Oriented Language—a higher level language used in business applications for payroll generation, inventory management, etc.

CONSOLE: a device that is used to control and monitor the operation of any computer system.

COURSEWARE: Curriculum-embedded software that includes computer programs, objectives, teacher guides, and supplementary instructional material.

CPU: Central processing unit—the section of the computer that is responsible for executing instructions retrieved from memory, including all arithmetic and manipulative operations. The CPU also controls input and output functions.

CRT TERMINAL: Cathode ray tube—an I/O device which uses a television-like screen for display generation (output) and a typewriter-like keyboard for input.

CURSOR: Indicator, often flashing, on a video display to designate where the next character input will appear on the screen.

DATA: Characters or numbers which are not part of a program of instructions. Data are used by a program to achieve arithmetic results or are output to or input from an I/O device.

DATA BASE: A collection of files of data, logically organized for efficient access and use.

DEBUG: To correct a program which does not achieve the intended objectives for which it was written.

DISC: A device inserted into a disc unit or drive on which programs and data are written in order to be safely stored for future use.

DISC DRIVE: An assembly of electronic and electromechanical devices that together achieve the objective of writing or reading programs and data to or from a disc. (also called disc unit)

DUAL DENSITY: Describes in relative terms the compactness of data written on a disc. Programs or data written on a disc in *single density* are less densely stored. More data or programs can be written on the same disc if stored in dual density than if stored in single density, provided that the disc itself will allow dual density recording.

EXECUTE: (run); cause instructions of a program, previously entered into a computer's memory, to each be acted upon with corresponding actions taken.

FILE: An organized collection of data stored in a computer (memory) or on a disc.

FLOPPY DISC: One form of disc. It is flexible, as opposed to some larger discs which are not flexible and encased in hard packs. A floppy disc is also known as a diskette.

FORTRAN: FORmula TRANslation—a higher level language developed for mathematical and scientific applications.

GRAPHICS: Pictures on a CRT, in addition to the simple display of letters, numbers, and special characters.

HARD COPY: Output produced on paper by a printer.

HARDWARE: The physical components of a computer system.

HIGHER LEVEL LANGUAGE: A method of computer programming which approaches the syntax of English and is easier both to use and to learn than machine language.

INFORMATION: The meaning associated with data.

INPUT DEVICE: A hardware component of a computer system which allows the user to enter program statements or data into a computer's memory.

INTEGRATED CIRCUIT: The plastic or ceramic body which contains a chip as well as the leads connecting the chip to other components or chips.

INTELLIGENT TERMINAL: A Programmable I/O device capable of limited processing functions independent of the computer to which it is connected.

INTERFACE: The electronic circuitry required in instances where the format or nature of communications between a computer and a peripheral device would otherwise be incompatible.

I/O: Input/Output—that subsystem of a computer system which embodies all peripheral devices responsible for inputting data to a computer or receiving the output of a computer.

K: Stands for "thousands" (actually 1024 in computer language) and is used in describing the size of a computer's memory or the rate at which data can be transferred from one point to another. As used in describing memory size, 48K, for example, actually means that the memory size is $48 \times 1024$ bytes or 49,152 bytes. 1K is equivalent to 1024 bytes of memory (2 to the 10th power).

KEYBOARD: Typewriter-like device which allows the user to input data or program statements into a computer.

LOAD: Transfer data or program statements from a peripheral storage device into the computer's memory.

LIBRARY: A collection of programs or routines which are readily accessible to a user. They are usually stored on a disc device.

MACHINE LANGUAGE: The language that allows instructions to be directly executable by the logic circuits of a CPU. The "menu" of these instructions is found in a reference manual.

MEMORY: That internal component of a computer system which stores instructions and data. It is organized with a certain number of storage locations. Each location, known as a byte, has a certain number of bits associated with it.

MICROCOMPUTER: A small computer, consisting of a microprocessor, memory, and associated I/O components.

MICROPROCESSOR: The CPU portion of a microcomputer. It resides on one integrated circuit.

MONITOR: A device which provides a video display.

NETWORK: A combination of computers and terminals linked together for communication purposes, as well as for the efficient use of peripheral resources.

OPERATING SYSTEM: A translation medium which allows the user to become divorced from the intricacies of machine language programming. The user can program in a higher level language, which is much easier to use and to learn.

OUTPUT DEVICE: Any component of a computer system which receives the output of the system. Examples include printers and CRT terminals.

PERIPHERAL: Any device external to the core of the computer itself, and which is used for purposes of facilitating I/O.

PILOT: Programmed Inquiry Learning Or Teaching—a programming language designed specifically to aid in the operation of CAI or CMI software.

PRINTER: A device on which hard copy is made.

PROGRAM: A series of instructions which, when executed by the CPU in sequence, perform work or achieve results.

PROMPT: A symbol appearing on a CRT display which prompts the user to input data or program statements.

SAVE: Transfer program statements or data from a computer's memory to a permanent storage facility such as a disc.

SOFTWARE: Instructions, or program statements in a higher level language, which are written, entered, and later executed by a programmer or user.

SYSTEM: The combination of hardware and software, as in computer system.

TERMINAL: A device which allows the user to communicate with a computer or the computer to communicate with the user.

## FEEDBACK FORM FOR READERS

Please complete the answers to questions asked below.

1. What portions of this text did you find most useful?

2. What portions of this text did you find least useful?

3. Related to question #2, why were the sections of the text you cited not useful? How could they be made more useful?

4. What terms, in your experience, were not included in the Glossary that should have been?

5. What computer-related concepts that you have come in contact with were not explained somewhere in the text that should have been?

Please forward completed feedback forms to Dave Drumm, c/o the publisher. Thank you!

# Index

GPSR Compliance

*The European Union's (EU) General Product Safety Regulation (GPSR) is a set of rules that requires consumer products to be safe and our obligations to ensure this.*

*If you have any concerns about our products, you can contact us on ProductSafety@springernature.com*

In case Publisher is established outside the EU, the EU authorized representative is:

Springer Nature Customer Service Center GmbH
Europaplatz 3
69115 Heidelberg, Germany

**Batch number: 10830937**

Printed by Printforce, the Netherlands